From Pessimism to Promise

From Pessimism to Promise

Lessons from the Global South on Designing Inclusive Tech

Payal Arora

HARPER
BUSINESS

An Imprint of HarperCollins *Publishers*

First published in India by Harper Business 2024
An imprint of HarperCollins *Publishers*
4th Floor, Tower A, Building No. 10, DLF Cyber City,
DLF Phase II, Gurugram, Haryana – 122002
www.harpercollins.co.in

2 4 6 8 10 9 7 5 3 1

P-ISBN: 978-93-5699-780-6
E-ISBN: 978-93-5699-957-2

Typeset in 11.5/15 Joanna MT Std at
Manipal Technologies Limited, Manipal

Printed and bound at
Thomson Press (India) Ltd

To my parents,
Rachna and Ranjiv Arora

The detailed notes pertaining to this book are available on the HarperCollins website. Scan this QR code to access the same.

Contents

Foreword

I HAVE SPENT THE PAST THIRTY YEARS MAKING A LIVING BY creatively solving other people's problems. I have had the privilege of reimagining how people work, play, learn and love. My work has taken me to baseball stadiums in the US, online dating apps in China, schools in Latin America and farms in Southeast Asia. In my role as managing director of Asia and global partner at IDEO, I'm constantly trying to make sense of the world around me and connect with others who can shine a light on what the future might bring. Payal Arora is at the top of my list when it comes to making sense of how to navigate many of these potential futures.

Payal and I spent the pandemic years debating and discussing where the world was heading and how we might do our part to shape it. Payal's critical eye and finely tuned sensibilities have made an extraordinary contribution to my worldview and the courage to solve new problems with new perspectives. Her book, *The Next Billion Users*, and its related TED Talk not only reinforced all that I had learned from working and living in China for fifteen years, but also helped to guide me in my more recent work across Southeast

Asia, Latin America and the Middle East. This book is the necessary companion to Payal's rigorous research and experience shared in *The Next Billion Users* and is for anyone interested in discovering why the Global South could shape our collective future in ways that could benefit us all.

One of the main challenges I encounter in conversations with Western executives leading global companies is the set of assumptions they make about markets in the Global South. Some see these markets strictly as sources of disproportionate growth, given their scale and emerging middle class. Even more alarming are those executives who are waiting for these markets to 'catch up' to the 'sophistication' of the Western consumer. Rare, though, is the executive who has realized the powerful contribution that the Global South can make to their innovation pipeline, their global purpose or their expanded and pluralistic worldview, and almost absent are those leaders who know how to build thriving businesses that fully realize the market's creative potential.

This book goes beyond simply highlighting how misunderstood the 'world majority' is in the West. Excitingly, it shows us how many of our *modern* challenges might be best addressed by proactively including the Global South in our digitally powered global future. By 'infusing our algorithms with the aspirations of the majority world', we can shape a more optimistic future for us all.

However, this book is not a feel good, do good, symbolic appeal—it is a rationally framed reality. The Global South, with its young and ambitious majority, approaches life with a 'do more with less' mentality. I see this approach alive and well in Indonesia's ministry of education, where they have built a substantive design team, sourced UX designers from the high-tech sector, who are rethinking digital learning services that will unlock the country's young population. I recently spent a week in Mexico City with IDEO's strategic partner Colectivo23, a digital skills academy for Latin America, where we met a wide range of practitioners, students and creators who are passionately developing the skills required to creatively drive the region's economies forward. These are the places where intent and action have come together to create tremendous impact under extraordinary constraints.

However, these kinds of solutions are not pervasive, nor are they a given. Payal describes the tensions that reside in trying to tackle some of our systemic challenges. How can we use centralized control to distribute access to opportunity? Why can't sustainable practices also stimulate growth? How could more global representation in our data deliver better digital outcomes for all? This book insightfully points out that we are trying to solve these challenges with technology when their root causes are fundamentally cultural.

Cultures Matter

Payal reminds us that culture is messy, contradictory, and not as binary as the algorithms we are so reliant on to increasingly run our world. Biases run deep, access to opportunity is restricted and a culture of fear seems to have replaced the culture of care we all so desperately need.

We have arrived at that juncture where we must decide between the scarce world of efficiency or the abundant world of creativity. To put it differently, will we pursue a path of least resistance or a path of most resilience? By addressing some of the cultural assumptions we are making about the Global South, we will discover technological win-wins for people and the planet, for the digital and the physical, for the West and the 'rest'.

In my fifteen years of living and working in China, I have spent a fair amount of time speaking to creative communities, teaching business and design students, and interviewing promising talent to work at IDEO.

I have, time and again, been impressed by the depth of awareness and understanding of global design trends and solutions these young Chinese designers and strategists have. They are both familiar with successful China-based innovations, business models and design solutions, as well as the equivalent examples and benchmarks from the West, from Southeast Asia and India, and from Japan and South Korea. This is in stark contrast to most of the young Western designers who have taken what happens in Silicon Valley to be the one and only 'global' frame of reference and who are not drawing from the abundant contrasting ideas that are emerging in the rest of the world.

As a designer and creative leader, I agree with this book's framing of many of the challenges of our time as opportunities for responsible design. While highlighting what is problematic about Western tech's current impact on the Global South, the book offers ways to design social solutions that can benefit all. From drones on African wildlife reserves to music-sharing in the Middle East to pornography as education in India, this book provides context and direction for business leaders, digital creators, policymakers and anyone else looking to realize a better future—a future that both benefits the Global South and is benefited by it.

I often say that many tech solutions are answers in search of questions. This book brings us back to those questions that matter and shines a light on the practices that we might use to move forward.

Ask Different Questions

We need to move beyond the paradigm of designing for 'good'. Whose good are we claiming? Morality-driven approaches are a misguided way of getting to the best outcomes. Payal proposes a shift from morality-driven design to relational-driven design, where we take a step back and try to understand the deeper layers of cause and effect at play, typically of human origin. For example, when trying to tackle the perpetual problem of slaughtering endangered animals on African wildlife reserves, Payal questions the focus of investment on ineffective high-tech solutions for tracking animals instead of trying to understand the motivations behind the culture of poaching. Another example given in the book is how the blind censorship of 'pornography' in its broadest definition fails to recognize how it is being used by young people as a surrogate for sex education. Why aren't we thinking harder about how we could take something that has been weaponized to hurt people and transform it into a tool for empowerment, intimacy and compassion?

The examples above highlight the importance of taking a critical look at the fundamental questions we should be asking, rather than racing to solutions. What are the underlying causes and dynamics at play that may be leading to the problem at hand? I cannot think of a better example than in healthcare innovation. The tremendous scientific breakthroughs that have created modern Western medicine and medical practice have neglected to integrate the far older wisdom of preventative well-being that comes from traditional Chinese medicine, Indian meditation or indigenous ecocentrism. One of my favourite activities when hosting a Western executive from the healthcare industry in Asia is to get them up at the crack of dawn and spend two or three hours in a park somewhere in China. They will see how the elderly ensure their later years are spent thriving, from dancing sessions to laughing clubs to Tai Chi to community debates—in nature, with others, and actively appreciating the present.

For all of its progress and benefits, the Industrial Revolution also separated us from the earth, compartmentalized our work into measurable productivity units, and put us on a path of mechanized efficiency. Digitization should not accelerate this linear path toward optimization, but rather, provide a circular rhythm that includes our cultural and environmental roots. By designing solutions imbued with the indigenous wisdom of ideas like frugality, the culture of collectives, subsistence and repair, we can deliver healing outcomes that sustain people and the planet. We must shift from the industrialist's description of data as the new oil to the more careful cultivation of data as the nutrients that fuel a regenerative and thriving ecosystem of interconnected living.

Human-Centred

As someone who has advocated for businesses and systems to be more customer-centred and societally inclusive, we must think harder about what human-centred design means and what it must do. We need to celebrate the boldness of human ingenuity while appreciating the humility of being *humane*. A healthy world is one where humans are not at the centre but are thriving participants, enabled by technology and grounded by nature. We must design for the tactile and the intimate, for the earthly and the cosmic, for the multicultural and the interconnected.

By opening our aperture on how we design and whom we serve to include the Global South, I am on board with Payal when she argues to tap into the ambitions and optimism of the world's majority—a majority that has proven time and again that it knows how to do more with less, generate solutions that improve our systems and possibly point us toward a renewed vitality in embracing sustainable growth.

As Payal shows, it is a moral imperative to design with hope. Some say that designers are natural optimists. Maybe that's why I've never been more excited to join the movement in designing what happens next, with the Global South at the front and centre.

Charles Hayes

Executive managing director, Asia; partner, IDEO

Introduction

I WAIT BACKSTAGE, WRAPPED IN A GIANT WOOLLEN SCARF. IT IS A cold September evening in Copenhagen. The 2019 Techfestival is set up like an open-air theatre in the Meatpacking district, surrounded by tents with disc jockeys, hackers, demos, podcasters and thought labs. There is a sizable crowd of 10,000 people registered to attend this three-day Techfest from all over the world.[1] I am about to give a keynote on the next billion users (NBU) in the Global South, optimistic about the digital future. Before me, the inventor of the hashtag, Chris Messina, delivers his talk.

He publicly apologizes for his invention. Messina has embarked on a talk circuit of penance that includes his interview for Netflix's globally acclaimed film *The Social Dilemma*, a dystopian docudrama that tells the story of how a group of tech elites controls the way 'billions of us think, act and live our lives'.[2] The documentary warns us that the technology that connects us also controls, manipulates, polarizes, divides and monetizes us. Messina echoes this in discussing the hashtag, dismayed by its viral toxicity. He promises to work on a solution to the problem: Frankenstein's monster in need of taming. His motto of the day is, 'You break it, you fix it.'

A friend of mine from New York called me right after she watched *The Social Dilemma* with her brother, anxious about tech. They felt helpless against what they learned about the constant surveillance of their everyday lives and wanted to know how to protect themselves from such digital control. My students in Rotterdam feel fatalistic about the prospect of having sovereignty over their data. Some peers at an artificial intelligence (AI) ethics workshop on good digital futures struggle to see the light as they lament how tech contributes to democratic decline. Policymakers are busy trying to break up, contain, regulate or ban social media. There is melancholy in the air as media scholar Geert Lovink puts it, 'The mobile has come dangerously close to our psychic bone, to the point where the two can no longer be separated. If only my phone could gently weep.'[3]

While legitimate concerns drive these fears, we need to account equally for young people's digital aspirations to ensure we do not inadvertently destroy that which they value the most—rare spaces for self-actualization. Anyone who has ventured outside the West will feel the contagious optimism from young people's extraordinary hunger for being digital. These youths are full of enthusiasm and passion for life despite the tremendous socio-political challenges they face. This is survival in its fullest sense. Finding humour while at war has helped people cope for millenia. Sharing something beautiful from a place of ugliness sustains human dignity. Creating moments of joy, inspiration and entertainment with digital media cultivates hope for the future. This dual perspective helps us reconcile the solidarities that come out of #MeToo and #BlackLivesMatter movements with current hashtag hating.

The West is suffering from pessimism paralysis—a negative bias towards all things digital—that can derail the drive to change the status quo. Negativity does not inspire change. Pessimism is for those who are privileged and can afford to live with despair. Imagine if Martin Luther King Jr., started his speech by saying, 'I have no dream.' It is not naive to be optimistic about our digital future. It is our moral imperative to design with hope.

Pessimism Paralysis

'How can billions get rid of something [social media] they do not even like, are addicted to, and do not know how to get rid of?' asks Geert Lovink, author of *Stuck on the Platform*.[4] Lovink and I debated Zoom fatigue, doom scrolling and other platform anxieties at the V2_, Lab for the Unstable Media in Rotterdam in fall 2022.[5] Geert is a self-acknowledged media pessimist, a sceptic of all things digital. The digital platform is out to trap us, design is out to depress us and the media is out to paralyse us, according to him. He is not alone in his bleak outlook on the digital future. On the contrary, his anti-establishment and dystopian stance is the norm among today's intelligentsia.

Internet scholar Safiya Umoja Noble argues in her 2018 book *Algorithms of Oppression* that 'algorithmic oppression is not just a glitch in the system but, rather, is fundamental to the operating system of the web'.[6] Her discovery of this inscribed digital oppression began with an innocuous search in 2011. She googled 'black girls' to look for activities for her nieces and got back hits like 'hotblackp***y.com'. The search triggered her to look deeper into the hidden values driving search engines. She argues that the platforms we use in our daily lives have oppression baked into their infrastructures.

Mainstream media agrees. The *Los Angeles Review of Books* recognizes that Google's algorithms are 'hard coded with white supremacy and misogyny'.[7] *Vogue* supports the idea that our algorithmic systems are 'predatory by design'.[8] Data scientist and whistleblower Frances Haugen made 'algorithm' a household term when she testified on

Capitol Hill in 2021.[9] She revealed how Meta intentionally boosts divisive content on users' timelines to capture their attention and foster depression. Public anger towards these systems is commonplace. Today, you can easily buy 'Fuck the algorithm' pins, buttons and stickers on Amazon.

What happens when your reality does not cooperate with this narrative? I had a junior researcher come to me to share her struggle. She was looking into how young people were being oppressed by social media in Vietnam and encountered an unexpected problem. Her participants were enthusiastic users of Instagram and TikTok, becoming even more so during the pandemic. They related experiencing joy, pleasure and support on these platforms. Her data would not fit the digital-oppression theory. When media outlets covered the internal reports leaked by Haugen, they chose to emphasize how Instagram makes US teens feel worse about themselves but neglected to share equally compelling evidence—that this photo-sharing app makes twice as many young people feel better about themselves.[10]

Why do we feel compelled to cancel the positive with the negative? Isn't contradiction at the heart of human nature? We fear what we love. Much like any good marriage, society's relationship with technology needs work. We need to draw boundaries, watch for imbalances, and ensure that the union stays on track. Being hopeful is fundamental to a joint future. Tech critics, however, would peg this attitude as missing the big picture, being too simplistic, and lacking critical judgment. Is tech optimism just naivete?

Nuance as Naivete

Pessimism is rewarding in the media-saturated world. Bad news is good news for clickbait.[11] More books are sold, talks watched and grants awarded with doomsday messages. Depressing headlines pull in readers. Experiments conducted on people's responses to negative news find that people react quicker to words like 'cancer,' 'bomb,' or 'war', than to words like 'baby', 'smile' or 'fun'.[12]

Devrupa Rakshit, associate editor of *The Swaddle*—a health, gender and culture news site—asks how people have come to 'associate negativity and hopelessness with a higher degree of intellectual aptitude'.[13] Rakshit surmises that pessimism is looked at as a call to action, while the opposite translates to happiness with the status quo. Cynicism is a measure of maturity and being critical, a sign of objectivity. British science writer Matt Ridley notes that if you say the world is getting better, you may be in danger of being called naive, insensitive or out of touch. On the other hand, Ridley remarks wryly, if you say that 'catastrophe is imminent', you may expect a MacArthur 'genius grant' or even the Nobel Peace Prize.[14]

It is even harder to speak with nuance about the Global South; ideological straitjackets confine us. Some of my Chinese academic peers have gotten increasingly frustrated as they cannot address the everyday social lives of their people without pegging it to Chinese authoritarianism. It's akin to asking scholars in the United States to always address gun violence, abortions and Trumpism when sharing insights on quotidian life in the United States. When I gave a talk on data governance to Silicon Valley folk, I remarked on how many Global South tech entrepreneurs admire Chinese innovators. Despite limited resources, China has been able to become a legitimate competitor to Silicon Valley. Chinese state support for e-commerce, mobile money and internet penetration over the last decade has resulted in unprecedented poverty reduction, becoming a model for other Global South countries to follow.[15] After the talk, I was asked if I would like to edit my part on China prior to the video release for the sake of my reputation.

A positive note on Chinese innovation does not translate to an endorsement of human rights violations. Intellectual honesty demands breaking out of binary cults. The nature of culture is contradiction. All communities straddle different generations, ideas, beliefs and experiences that make for a rich and diverse social history. Sensational declarations need to give way to sensitization to stories. Courage is often mistaken for naivete. Is it foolish for young women to take to the streets to defy the ayatollahs in Iran against the mandatory Islamic dress, sometimes at the cost of their lives? Were Ukrainians irrational to take on Russian forces when they were invaded? Should we consider young people ignorant as they take to social media to express themselves online, despite knowing that they are being watched, heard, monetized and regulated? Hope comes when aspirational forces overthrow oppressive regimes, causing perhaps the right dose of algorithmic anarchy for social change.

Algorithms of Aspiration

The concepts of leisure and poverty sit uncomfortably together.[16] It feels perverse to position a hammock against hardship. I have offended practitioners and policymakers in the aid sector, who view the coupling of these two notions as a mockery of the margins. But it is a celebration of human resilience, creativity and ability to find pleasure, opportunity and solidarity, despite adversity.

Global citizenship surveys over the decade capture a striking fact: while young people in the West feel dismal about the future, those outside the Western bubble are optimistic. A 2016 poll from the BBC suggests that while a majority of youth in Nigeria, Peru, China and India see themselves as global citizens and are enthusiastic about their place in the world, those in industrialized nations feel the opposite.[17] In Germany, for example, only 30 per cent of respondents see themselves as global citizens.

The Varkey Foundation did a survey in 2017 across twenty countries among Gen Z, those born between 1995 and 2001, on their perspectives of well-being.[18] China and India have the highest percentage of young people who believe that the world is becoming a better place. In Europe, especially in France and Italy, young people feel that their future is bleak. The report finds a clear division 'between the optimism of the developing world and the pessimism of the developed world'.[19] Despite these attitudinal differences between the Global North and South, a 2022 UNESCO report reveals that youths, across contexts, are committed to building their knowledge and engaging with technology to shape their lives.[20] There is consensus among hundreds of young people from forty-three countries that digital access and usage are the driving forces for a hopeful future, despite growing digital surveillance in their countries.

Current algorithms may be rife with oppressive, racist, misogynistic and discriminatory values. Yet there is power in numbers, as users from the majority world infuse these algorithms with their aspirations through their everyday digital acts. Small acts in conjunction can create seismic change, creating a new digital order.

Opting In

During the pandemic, the Innovation Services division of the United Nations Refugee Agency commissioned me to study digital leisure among Venezuelan refugees in Brazil. They had read my book *The Next Billion Users*[21] and took seriously my critique about how aid agencies approach their beneficiaries through a utility-driven lens. They made digital leisure the theme of their 2021 work plan. My team looked at refugees' digital immersions and the impacts on their everyday lives—their relationships, their prospects, their hopes and aspirations, and their sense of belonging.[22]

We found that far from wanting to get off of tech platforms, refugees wanted in—desperately so. Connectivity was gold; it was traded like a precious commodity. Data top-ups, Wi-Fi passwords and access to mobile devices accorded status and currency. Facebook and WhatsApp uplifted many of the youth. Hugo, a nineteen-year-old homosexual Venezuelan man, keeps his spirits up by sharing his vulnerability on Facebook. Doing so allows him to be himself, gain support and develop a sense of self-love after much persecution. He is particularly proud of one of his posts:

> A picture I took expressing that I felt very lonely. That picture made it to the top 20, people loved it. I liked that picture because it was the only one in which I felt I valued myself. It was with that picture that I understood that I have to value myself as a person.[23]

Lucía, a thirty-year-old woman, travelled to Brazil due to health issues and the lack of medical access in Venezuela. She suffered from endometriosis and needed surgery. She enjoys watching funny videos and religious content that promotes reflection, and wants to tell her story online as she believes it will give others like herself strength and spiritual support when they realize they are not alone.

> My dream is to make videos because my story is hard to tell. I went through five surgeries and God lifted me up. Many people are in the same situation. I had two heart attacks and God raised me up and many things happened … There are things I want the world to understand.[24]

Meta has lost face, but not lost ground, especially among the new generation in the Global South that has come online. WhatsApp, Messenger, TikTok and Instagram rule the hearts and minds of many young people from Africa and India. During the pandemic, these apps became the main venues for self-expression, social belonging and livelihood when public spaces were restricted overnight and street vendors were forced to go online.

Youths have come to terms with the idea that their future lies in their hands. Their aspirations are not aligned with what their governments have in store for them when it comes to future work opportunities, and they are determined to not reproduce their families' pasts. These youngsters will not be farmers, maids or street sellers in a traditional sense. Digital media is often their way out of their confined contexts. They reinvent the game in line with their dreams, dedicating themselves to cracking the code and hacking creativity online. They face oppressive laws, patriarchal cultural norms and restrictive social rules, and they are not naive about the political and social harms that come with being digital. Relative to these static and confining conditions, digital media appears more dynamic and forgiving as it allows for more experimentation. The algorithm, despite the risks, is still their friend.

Next Billion Users Labs

Tech companies' interest in the Global South since 2015 is no surprise. In 2017 *The Economist* published a story that asserted 'the world's most valuable resource is no longer oil, but data'.[25] Companies are in an AI race. In this data economy, the one with the most data wins. Companies have witnessed how millions of users, primarily from the Global South, are coming online for the first time. India and China alone are home to most users today, and neither market has reached saturation.[26] Both countries radically disrupted the mobile and data industries by making their devices and data plans extraordinarily cheap and user-friendly. This has enabled young people with limited financial means to go online and participate in the digital economy and social media world for the first time. While 3.8 billion people—half the planet's population—are low income, they are enthusiastic adopters of the mobile phone and the internet.[27] From South Africa to Brazil, young people are taking to the mobile internet with gusto.

Around when *The Next Billion Users* was released in 2019, Spotify and Intel launched their next billion users labs to study users outside the West. Google grew their NBU lab exponentially in 2020 from a few dozen team members to hundreds.[28] Meta's emerging markets group made way for their next product experimentation team. Launched in 2019, this team intended to test the new hypothesis 'that the next big idea may come from a market outside the US'.[29]

From connecting the next billion users, companies are now analysing their data in ways that can add to their organization's competitiveness. Western innovators are not just trying to capture the NBU market, they are trying to copy it. In 2019, Amazon launched its livestreaming for retail but admitted that it was not novel. By then, China's Taobao livestreaming network was already operating profitably, with their top influencers generating $15 million in sales through this e-commerce platform.[30] A major reason for Taobao's success is that it treats livestreaming as the mainstream e-commerce model and focuses on lower-tier cities, where influencers have made about 70 per cent of their sales.[31] In 2023, while Amazon continues to struggle to make a success of their livestreaming portals, Taobao has become one of the largest consumer-to-consumer (C2C) e-commerce platforms in China and has an estimated monthly traffic of 240 million users.[32]

Despite Western tech's shift in focus towards the Global South, deep stereotypes prevail. Companies continue to view lower-tier cities and income groups as high-risk investments. Funding agencies view these populations as beneficiaries, not collaborators. Most scholars are focused on the oppressive side of tech, leaving little room to investigate the other side of the coin. This results in weak business models, inaccessible products, misguided or imbalanced ideologies, and obstructive policies.

The Global South by Any Other Name

'Rest of World' (RoW) is a corporate term still in use among Western businesses when they talk about 'everyone else' outside the West.[33] This lumping together of disparate cultures and populations has a legacy. During the Cold War, the North Atlantic Treaty Organization (NATO) and its allies labelled countries in Asia, Africa, Latin America and Oceania that were economically poor and non-industrialized as 'third world' and 'developing' nations. These political categories mirrored postcolonial relations with and placement among the 'first world' and 'developed' nations such as the United States, Canada, Australia and European countries.

In the 1980s, the International Finance Corporation did an economic makeover of these populations. They called them 'emerging markets' to attract investors.[34] In the 1990s, aid agencies and policymakers embraced the term 'Global South' for countries outside the West. This term reflected a shift in focus from geographic, developmental or cultural differences, towards geopolitical power relations.[35] This term remains the preferred catchphrase to describe regions with shared sociopolitical struggles and cultural histories despite its flattening potential.

In the 1990s, the late business guru C.K. Prahalad invigorated the business world to rethink low-income groups in countries like India and China as legitimate markets due to their economies of scale. He termed them 'bottom of the pyramid' (BoP) markets that required new approaches for win-win solutions of doing good and making profits.[36] Today the next billion users are the new BoP in the digital economy. Yet it would be a shame to view young people in the Global South as mere data points that need extracting to build tools and services for the West. They are creators, entrepreneurs and innovators forging ahead with new ways to think and be online.

Figure I.1. How these terms evolved

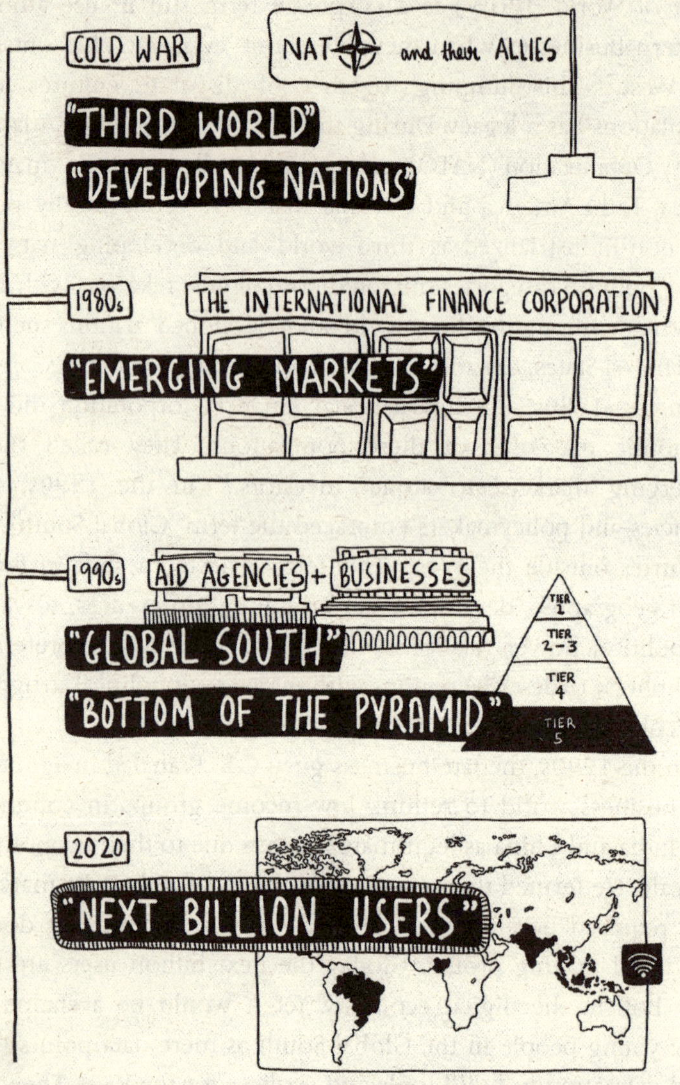

Old Mental Models Die Hard

Since the release of *The Next Billion Users*, I have given talks at several US and European tech companies and venture capital firms on the rise of this market. My mission is simple: to convince Global North organizations to value people in the Global South as legitimate consumers, creators and citizens. It is an uphill battle.

In one of my consulting gigs with a Western tech company, I had just finished my presentation on the dynamism and creativity in the next billion group. This was followed by someone from senior management saying wryly, 'If we don't know what the business model is here, all this can get incredibly boring in a very short time.' It is clear how little time and patience is dedicated to understanding and learning from regional innovations in the Global South.

On the flip side, many Western academics feel disdain towards all things market-based. Monetization is morally repugnant. Capitalism is a dirty word. Neo-Marxism is alive and kicking in Western academia. Critical media scholar Christian Fuchs pushes for a radical change from market systems as 'capitalism has the potential to produce Auschwitz'.[37] Framing youth online in the Global South as under-explored markets has become, as media sociologists Nick Couldry and Ulises Mejias argue, 'capitalization of life without limit'.[38] These scholars are understandably tired of users being pursued in order to harvest their data and want to build citizen control, choice and confidence for 'data justice'.[39] Fair enough.

That said, youth from resource-constrained contexts have a radically different starting point, having been ignored by markets for decades.[40] They have had to hunt hard for good deals, products and services that cater to them. Companies offer people living in remote, rural regions few choices, resulting in higher prices for the same goods and services than their urban counterparts enjoy. This 'poverty tax'[41] is imposed on those already at the margins. In contrast, Western consumers remain at the centre of business models, tech services and future-shaping policies, constantly sought after by companies. For the Western customer, a life without the market is unimaginable. In framing young people in the Global South as the next billion users 'market', I am beseeching companies to treat these users with respect as legitimate and viable consumers. Once this baseline is established, these users can exercise their consumer rights in the digital and global marketplace.

The Long Tail Promise

Silicon Valley needs more convincing that Global South users constitute worthwhile markets; Western academia digs deeper into anti-market dogma, decrying data colonialism. Meanwhile several global tech entrepreneurs and innovators are increasingly interested in reaching youths in the Global South in ways that are more nuanced and far-reaching.

The CEO[42] of a global seed foundation expressed to me his frustration about the low uptake of his AI-driven agricultural products by the youth in the Global South, as they remain disinterested in being farmers. He sees the value in content creators opening a dialogue between agri-tech businesses and rural youth to counter this disinterest and spur uptake among smallholder farmers (an estimated 500 million worldwide) in the years to come. Traditional training programs will not cut it. What he needs is 'high involvement with the rural poor in emerging countries'[43] and outreach with the next billion users through digital means. 'We're not just ticking boxes and crunching numbers,' the CEO explains. He believes that the agricultural sector data, when used correctly, can be a game-changer in providing entrepreneurial opportunities and new livelihoods, making the sector economically lucrative.

Similarly, leaders of a solar energy company sought to partner with me to build an outreach package for rural households in Africa. They were concerned that solar panels were too abstract and the company did not have the pull to get Africans on board alone. They envisioned a 'package' of a solar home system tied to a family smartphone plan, competitive data bundles, and most importantly, local content creators driving engagement with an understanding of new products and services.

Entrepreneurs seeing value in southern creators is a pivotal shift. Local content creators in the Global South are gaining credence among global startups, and are serving as essential liaisons between the market, states and end users.

AI in the Wild

AI, like every other technology, is only truly understood when played out in the 'wild'.[44] As the digital becomes integrated into our everyday lives, it becomes a social space, and a pathway to explore our identities, relationships and cultures.

The in-the-wild approach tests products and services in real-world settings with different groups and cultures. While user-centred design (UCD), which targets users' needs and preferences, has been around for decades, little attention has been paid to users in the Global South. Even today many tech companies use their outsourced hubs in India or Indonesia for doing back-end work for users in the Global North. Several US companies I've consulted for have their southern research teams stay up late to do user-research on consumers in the United States.

Computer scientist Bettina Berendt reminds designers that when we design AI tools, we are not solving chess problems. Most social problems, she argues, are driven by ideologies with 'certain versions of the notion of the Common Good'.[45] People in power often define what is good and bad. Look at the cast of characters that boost 'AI for Good' projects—they are almost all from the Global North. These do-gooders need to offer grand visions to get institutional grants, pass global policies and gain messiah-like status in their domains. It doesn't pay for tech innovators to focus on AI for effective governance and accountability. Tech evangelists are allergic to the boring stuff that most of the world clamours for—getting the basics to work. Instead, they pioneer new apps, gadgets, systems and platforms. Tech experts resort to tech solutions for tech problems; their solutions demand the lab over the wild. Worse, they may treat the wild as their lab, dehumanizing people by turning them into mere data for profit.

Figure I.2. Global South as a live laboratory

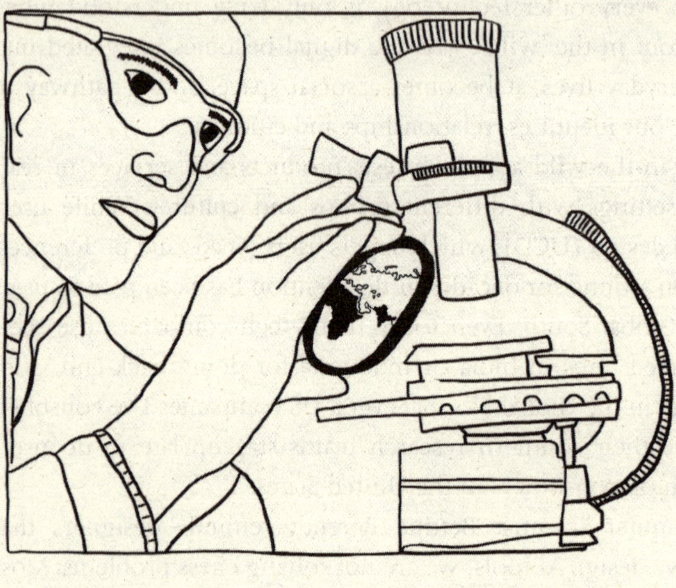

In recent years, the fields of user experience (UX) and digital ethnography have expanded. UX researchers serve as a counterforce to the laboratory approach as businesses wake up to the social side of the digital. Companies are slowly recognizing that when we ask ourselves what the future of tech will be, we are inadvertently asking ourselves about the future of human creativity, empathy, fairness and care.

AI is not just a tool; it is an ideology.[46] Digital ethnography can reveal the values that shape these ideologies. As AI unpacks human culture, UX researchers can, in turn, unpack the culture of AI. After all, humans are woven into the entirety of the AI supply chain—the data fuel that keeps the algorithmic loops spinning. They are the data feeders, framers and forgers of the AI worlds they inhabit. You don't need to be an engineer to crack the code. UX experts are capturing how young people at the margins are making metrics matter for themselves by gaming these codes of practice to become visible and heard. These youths, often missed by mainstream media, are determined to tell their stories by pushing creative content ignored by the traditional creative industry.

UX experts can be powerful allies as they observe, document, translate and amplify these stories. Yet, qualitative work and its inherent messiness needs constant championing against age-old decision-making practices that privilege statistics over stories.

In Defence of the Anecdote

The word 'anecdote' comes from the Greek term 'anekdota', which means something unpublished. Anecdotes can bring stories alive for decision-makers and nudge them to prioritize underserved communities. Many of these stories are not just about the past and present, but also about people's aspirational futures. While aid agencies are undergoing introspection and signalling humility about their role in social change, policymaking stubbornly remains driven by metrics.

I was on a panel with leaders from Oxfam Great Britain, the African Visionary Fund, the United Nations Development Programme (UNDP), the Sierra Leonean Ministry of Education and the Indonesian Ministry of Finance for the Istanbul Innovation Days 2021, a UNDP event. Its byline for development futures was 'What's after the WESTERN model of development?[47] It was refreshing to share a common starting point with my panellists, who also sought alternatives to existing development models. The CEO of Oxfam Great Britain, Danny Sriskandarajah, argued that we need to build from below a 'twenty-first century network' that is diverse and globally balanced, yet keep the 'audacity to be internationalists', by recognizing our collective well-being.

I shared insights from two years of fieldwork during the pandemic on women workers' experiences with gig platforms in India. An audience member asked me how I could expect such 'anecdotes' to help in decision-making. This wasn't a surprising question; anthropologists in policy and business circles are often asked it. Studying culture is messy, difficult and not computable. If done right, it produces a diversity of narratives, values and perspectives that goes to the heart of human motivation. But institutions love statistics: statistics are neat, simple and give the impression of being objective. They are also replication friendly. Big data analytics abdicates decision-makers from getting emotionally involved. However, decisions are rarely a numbers game.

Whose Knowledge,[48] a feminist collective that aims to decolonize knowledge, makes the case that when building movements online, organizations tend to focus on what can be counted—the 'transactional metrics'.[49] The collective goes on to argue that in doing so we miss out on an equally important measure of how relationships and collaborations establish and institute change when they go online—what they call, 'transformational metrics'.[50]

Behavioural economist Dan Ariely explains that institutions tend to get what they measure, so they measure what they want. People adjust their behaviour based on the metrics they're up against.[51] This skews an institution's perception of what is to be resolved. Anecdotes come together to make compelling stories. Stories come together to excavate the unseen and the unheard. Culture, while complex, can be studied and mapped by engaging with diverse people as they share their experiences with new technologies. Users' practices, while not static, remain relatively stable, and make sense when tethered to their social environments. While not homogenous, digital culture is shared between generations.

Anecdotes change over time. What works shifts based on what people value at a given point in time and space and in relation to those around them. These values shape the rules of the game. It takes a unique combination of collective drive and restraint to maintain the borders of user practice. Within these borders new stories are told.

My Approach

All chapters in this book contribute to new ways of seeing and doing that include the majority world. These texts push against the pessimism pervading our social and digital lives with the counterforce of the aspirational usage of youth in the Global South. My arguments reveal incongruencies between the stated intent and realities of the impacts these tools have on people's lives. Being optimistic about the future doesn't negate critiquing existing policies and practices.

I draw my insights from the hundreds of engagements I have had since 2020—with practitioners, programmers, tech experts, development leaders, innovators, activists, venture capitalists and non-profits through consultancies, keynotes and workshops. I have spoken to philosophers, corporate leaders, computer scientists, designers, artists, economists and media anthropologists over these years. In this process, I gained much respect for the labour that goes into building a common vocabulary and shared starting points. In my consultancies with tech companies, I have had to speak to members of diverse teams separately. I engaged with expert groups in equity products, international design, cloud computing, marketing, education and business development, who often had little crossover and few internal synergies with one another. Much like governments, corporations struggle to come together for joint missions as they have their own agendas, budgets and politics.

In this book, I have tried to bridge the user experiences of young people at the margins of society with institutions and policymakers at the local and global levels. Within these tensions and negotiations, I offer ways to move forward and think of what global digital well-being and inclusive design can look like.

What to Expect

Chapter 1, 'AI for Good, Bad and Ugly', focuses on the growing number of AI for Good projects and underlines the need to shift from morality to relationally driven approaches to design. This pivot from altruism entails investing in how relations between people, places and technologies can work in ways that sustain human dignity, civility and responsibility. The text draws from fieldwork on AI for Good projects such as the deployment of AI-enabled anti-poaching tools in Africa to reconfigure the ways we account for doing good.

Chapter 2, 'Algorithms of Aspiration', offers ways to rethink digital creativity and the creator economy that is increasingly engaging the new generation. This chapter addresses one of the most urgent questions of our day regarding the rise of creative AI technologies: who and what counts as creative in this AI-driven era? The chapter draws from an extensive UX consultancy that my team undertook in India from 2021 to 2022 for a major design company. We examined how young creators across socioeconomic groups use digital tools and platforms to create content, manage their data, develop creative processes, access tools, and express themselves in order to build community and monetize content. In this chapter, I push back against conventional notions of creativity, which have long been dictated by the aesthetic tastes, values, needs, concerns and aspirations of the West: a practice that has often resulted in suboptimal or unusable designs for the majority world. I also tackle questions on creative rights, provenance, fairness, labour and representation.

Chapter 3, 'Digital Desire', argues for reimagining pleasure-based perspectives on design. This approach is a response to growing loneliness, and the need for intimacy, pleasure and play with our sexual selves. Most of the world holds conservative and patriarchal values, and restricts sex and sexuality. Going against that trend, this chapter captures the spectrum of tactics, and the extreme lengths young people will go to access sexual content, romance and intimate relations. This is made visible in back-end metrics of digital campaigns on sex and pleasure in Egypt, Kenya and Mexico, or where Pornhub becomes sex educator to the young generations. Self-care apps, the fastest growing genre of apps, takes on stealthy new forms in the Global South in response to the urgency to move from the war on vulgarity to making peace with sex.

Chapter 4, 'Surveillance of Care', is about reclaiming surveillance' as a positive in social well-being by inserting the dimension of care. The term 'surveillance' causes anxiety today as a concept that is associated with corporate tracking and oppressive government control. We seek to limit it, curtail it, disrupt it. However, not all forms of digital surveillance are bad. Surveillance care systems have long served as a positive force against infectious diseases, and are used for eldercare and self-care. We need to scale that logic across our digital lives, taking action that is preventative, nurturing and cooperative. This chapter taps into UX research on lessons learned from feminist digital activists and women's collectives in the Global South as they craft ways to replace commodification with care in their navigations online.

Chapter 5, 'Green Design', makes the case for indigenous approaches to design as a key pathway to a sustainable future. Indigenous cultures have long valued frugality, repair, collective action and subsistence, many of which are appropriated by popular Western models of sustainability today. This chapter helps to reposition the models of degrowth, circular economy and the doughnut economy in a larger paradigm of user choice versus social condition. With data becoming the new oil, data is also a material reality. The rise of the next billion users causes panic as digital engagements demand sizable pieces of the energy pie. I delve into historical practices of consumption and identify current values that inadvertently pit the digital inclusion of the Global South against climate justice. This chapter reframes global resiliency, regeneration and redistribution by re-aligning Global South realities with northern frictions of past practice.

The thread running through these chapters is the growing interest in the next billion users and its implications on design thinking and practice. I advocate revisiting traditional businesses, typical products, existing policies and design models in response to this phenomenon. The reverse engineering of our systems demands a reverse flow of wisdom from the South to the North.

As AI disrupts sectors across the board—creative, pharma, security, agriculture, education and beyond—people in the Global South can shine a light on navigating these disruptions as they have had to learn to reinvent themselves quickly, leapfrogging obstructive legacy systems, norms and practices, and moving towards a positive future.

1

AI for Good, Bad and Ugly

IMAGINE A WORLD WITH NO ELEPHANTS. THIS FICTION COULD become fact in a mere decade, as poachers kill an elephant every fifteen minutes.[1] Rhinos, tigers, gorillas and other large mammals also face extinction due to illegal poaching across national parks in Africa and Southeast Asia.

The conservation community considers AI to be one of the most important game-changers in doing good.[2] According to a 2021 report[3] from WildLabs, a central hub for conservation technology, AI can reduce the physical labour that goes into collecting vital conservation data. AI can pinpoint an animal call from hours of field recordings in mere seconds and identify rare species from thousands of images online. Where humans fail, AI may be able to step in and save our planet.

But what makes AI 'good'? Is it sufficient to have noble intent when designing AI-enabled tools? Are they tackling the most important problems? Do AI stakeholders agree on what needs fixing? What is the nature of the data feed that informs policy and practice?

Garbage in means garbage out. Many experts today are attending to digital rights, data justice and AI ethics to correct AI biases in the Global North. When shifting contexts to the Global South, however, tech philanthropists, AI evangelists and utopian gurus lead the way. AI for Good initiatives are the latest manifestations of their supposed altruistic efforts.

In the ping-pong game of good and evil, there is little room for nuanced values. With grand ideologies, the mundane gets left behind—the stuff that dictates our digital and social lives. To stay on track to make a positive social impact using AI, reliance on morality needs to give way to relational approaches to tech innovations. Taking a deep dive into the culture on the ground in the Global South and affording fair weight to people's lived realities there will help AI find a home in the hearts and minds of those most affected.

When AI Needs the Pope's Blessings

The Vatican hosted an unusual guest in February 2019: Microsoft president Brad Smith. Pope Francis, the Roman Catholic leader, who, some years earlier, remarked that he was a 'disaster' when it came to technology, joined Brad Smith to discuss AI for common good.[4] In the last decade, AI for Good has grown as a field, as an ideology and as a mission to save the world. With no country on target to meet the Sustainable Development Goals (SDGs) by 2030, the UN has doubled down on AI to accelerate progress.

Doing good has always been a big business. With AI, it is on steroids. Fei-Fei Li, former chief scientist at Google AI, placed her bets on AI when she declared, 'I believe AI and its benefits have no borders … it has the potential to make everyone's life better for the entire world.'[5] She is not alone in this exhausting clarity on AI-enabled solutionism. The breeding season of non-profits for AI for Good kick-started a decade ago. For instance, in 2015, the AI for

Good non-profit was established to find solutions to advance the UN's sustainability mission.[6] Its headquarters is, unsurprisingly, in Berkeley, California, where it sits near others in Silicon Valley to innovate for good in the West to trickle down to the rest. The usual tech suspects have joined in, propelling a new generation of lab geeks toward this mission. Google has channeled $25 million to support the AI for the Global Goals project.[7]

What happens in Silicon Valley does not stay in Silicon Valley. There is a cascading effect on the usage of limited resources, energy and drive for AI in countries like Nigeria and India. Tejumade Afonja, a twenty-four year old Nigerian with engineering training, believes, 'We [Nigerians] could compete … AI is something Africa could do along with everyone else.'[8] Africa is touted as AI's future, given it is the world's youngest continent. Its population has a median age of twenty and 60 per cent of residents are under the age of twenty-five.[9]

In the meantime, Indian tech gurus are busy Indianizing AI. The 'prophet of inclusive innovation' R.A. Mashelkar calls it 'Gandhian engineering'.[10] AI can bring to fruition Gandhi's thinking on science where 'there is enough for everyone's needs, but not for everyone's greed'. He proposes the acronym ASSURED to illustrate seven key factors to take it the Indian way: affordable, scalable, sustainable, universal, rapid, excellent and distinctive. These are the new converts to the missionary zeal for AI.

With God in the picture, the devil is not far behind. Chris Nodder, author of *Evil by Design*, paints a dire scenario of how design can lead us to the dark side.[11] Our seven deadly sins get a tech makeover. Pride translates to products pandering to user values. Sloth builds a pathway of least resistance, as with the infinite and mind-numbing scroll. Gluttony taps into customers' fear of missing out to keep them connected. Anger finds its home in anonymity, channelling the world's angst through digital taunts and trolls. Envy is all about status feeding into user aspirations. Lust nudges users to trust emotion over rationality; and greed reinforces customers' behaviours to keep them perennially engaged.

Faith drives the way we imagine tech innovations. When God speaks in the tech cathedral, the bazaar is forced to listen.

When Good Was Bad

News outlets are increasingly covering issues of algorithmic discrimination, predictive analytics and automation, and exploring how they infiltrate welfare communities, policing, healthcare and surveillance of public spaces. Public officers face some soul-searching today as they straddle AI-enabled efficiency and social justice. Aid agencies and state actors stand on a graveyard of failures of global innovations that claimed good intent—for instance, the last fifty years of trickle-down economics that justified tax cuts for the rich purportedly to benefit the poor.[12]

Few, particularly in Latin America, can forget the crusader Jeffrey Sachs with his 'disaster capitalism'[13] in the 1970s. He justified administering 'shock therapy' to the world's economies outside the West as a means for unparalleled economic prosperity for all. Today, tech saviours offer their AI-enabled medicine packaged as goodness for all to swallow. AI is meant to aid in curing economic recession, geopolitical wars and planetary degradation.

Doing good comes with a toxic legacy of oppression in the name of upliftment. The centuries-long colonial project was the most ambitious in its agenda for doing good. The English language was one of the most potent instruments to shape the 'good subjects'. During the British Raj, Lord Macaulay, the secretary of war in colonial India, became infamous for 'Macaulay's Minute' in 1835 where he focused on the deficiencies of the natives' cultures and languages, and proposed the English language as a solution. He argued that it was an 'indisputable fact' that Asian sciences and languages were insufficient for advancing knowledge. Only English could truly give us the 'enlightenment of our minds' and unravel the 'richest treasures of modern thought and knowledge'.[14]

'Macaulay's Minute' was an 'English for Good' moment of the day. English continues to enjoy a privileged place in the former British colonies, and is equated with progress, mobility and modernity.[15] Historically, what qualified as good was often bad, as it translated to subservience, oppression and taming 'subjects' to the existing social order. Even today, 'Uncle Tom' is a slur among African-American communities as it has come to mean a 'good slave' who complies with his master at his own expense.[16]

A 'good woman' of virtue and honour remains a live template in patriarchal societies, though many continue to resist it. A case in point is the Iranian women's uprising, defying the morality police as they burn their hijabs. A 'good subject' for governance under colonial rule was racialized and legitimized by philosophers, who are still revered and often uncritically taught in our institutions. Philosopher Immanuel Kant's hierarchy of 'good subjects' was based on supposedly innate traits of entire groups. For instance, he asserted that Indigenous Americans, as a race, were 'incapable of any culture' as they were too weak for hard labour and industrious work.[17] Old intellectual habits die hard.

Moral awakenings do happen, however. They have served as land mines in ideologies of goodness, blowing up atomized and culturally myopic constructs. Today's intelligentsia has come quite a distance from celebrating the nineteenth century poet-novelist Rudyard Kipling to condemning him. His work fostered empathy for the white man for the 'burden' he must carry to transform colonial natives from 'half devil and half child' to 'good subjects', worthy of compassion.[18]

Today, tech companies hungry to capture the next billion users market, frame their presence in the Global South in altruistic terms.

AI for Good projects driven by tech philanthropy spread the narrative that tech can solve the problems of 'the natives' by tapping into their aspirational yearnings for freedom, stability and inclusion. History does not die; it is reborn. Sometimes, in the best of ways, through human reckoning or the resuscitation of beautiful ideas of the past. Sometimes in the worst of ways, a nostalgia for the 'good old days' with even a fervent zeal to re-establish moral clarity with a binary logic of good and bad. To ensure that computational cultures automate the good and track the bad, we need to heed the diversity of contexts and peoples they strive to impact.

The Bretton Woods of AI

In October 2019, the Rockefeller Foundation brought together a group of technologists, philosophers, economists, lawyers, artists and philanthropists at their Bellagio Center in Italy to explore how AI can 'create a better future for humanity'.[19] They called it 'The Bretton Woods for AI: Ensuring Benefits for Everyone'.[20]

Seventy-five years ago, the Bretton Woods Conference laid the foundation for a global economic order. Nobody wanted another world war. The summit called for a radical reimagination of the financial system, trade and markets at a global scale. Institutional invention led to the birth of the International Monetary Fund (IMF) and the World Bank. The vision went beyond the economic paradigm with the birth of the UN, attending to how social and cultural factors shape global governance.

While these efforts contributed to decades of peace and prosperity in many parts of the world, they also led to rising inequality among nations.[21] Governance, over the decades, mirrored colonial relations. Europe and the United States shaped the leadership of these global organizations. They fixed the agendas and laid out the priorities for others to follow. Times have changed today as post-colonial nations have newfound confidence. In this data economy, the Global South demands data sovereignty. Global South countries have the world's majority of online users and data as well as indigenous companies that can build infrastructures to protect and preserve their nations from foreign extraction.[22]

The Belt and Road Initiative (BRI) financed by China has built infrastructures and alliances between China and East Asia, Africa and Europe. This geopolitical pack is already having a significant impact on AI-driven alliances and agendas, leaving the old guard from the United States and Europe outside the decision-making.[23] The French finance minister, Bruno Le Maire, admits, 'The Bretton Woods order as we know it has reached its limits.'[24] He makes an urgent case to reinvent Bretton Woods by acknowledging shifts in geopolitical power, else the 'Chinese standards—on state aid, on access to public procurements, on intellectual property—could become the new global standards'.[25]

The Rockefeller event was one such effort to reinvent Bretton Woods. I was invited to be part of this group alongside thought leaders like Rumman Chowdhury, then director for the Machine Learning Ethics, Transparency and Accountability (META) team at X (formerly known as Twitter); Joy Buolamwini, computer scientist and founder of the Algorithmic Justice League; Tim O'Reilly, founder of O'Reilly Media; economist Mariana Mazzucato; Ronaldo Lemos, director of Institute of Technology and Society in Rio de Janeiro and others. Many in this group, while diverse in their leadership, expertise and organizational affiliations, were pegged to the Anglo-Saxon context.

While others spoke about the data deluge, data glut and the need for data privacy, Ronaldo Lemos and I spoke about data blackholes, data deficits and the urgency of data visibility. In an age where visibility generates engagement, what does not exist online, does not exist in our social imagination. Informal settlements from Chad to Brazil, where more than a billion people live in precarious arrangements, show up as blank areas in location-based apps. Data deficits create and amplify marginalization for millions of people, as in the case of Brazil, where 1.5 million people live in informal housing—the favelas. Ronaldo Lemos explains that by being unmapped, people living in the favelas have limited access to public services. 'They don't have an address, so they don't get mail at the post office ... You don't get garbage collection, you don't get electricity. That's the reality.'[26]

Demand for Data Presence

Digital absence negates the lives of billions of people already at the margins. Object-recognition algorithms perform significantly worse at identifying household items from low-income countries compared to items from high-income countries.[27] Researchers at Meta's AI Lab tested some common household items like soap, shoes and sofas with image-recognition algorithms by Microsoft Azure, Clarifai, Google Cloud Vision, Amazon Rekognition and IBM Watson. They found the algorithms were 15–20 per cent better at identifying household objects from the United States compared to items from low-income households in Burkina Faso and Somalia.

Figure 1.1. Visually biased datasets

In less wealthy regions, a bar of soap is a typical image for 'soap', while a container of liquid is the default image used in wealthier counterparts. Data feeds for 'global datasets' for basic commodities come primarily from Europe and North America. Africa, India, China and Southeast Asia are the most under-sampled for visual training data.[28] The fact is that some groups enjoy more empathy than others due to long-standing efforts to make them visible and understandable. The prototype user for the designs of our everyday lives has long been 'WEIRD'—Western, Educated, Industrialized, Rich and Democratic.[29] The rest of the world has had to contend with centuries of colonial and patriarchal baggage, which have resulted in cultural erasures and historical fictions in mainstream narratives.

In *The Next Billion Users*, I explored how the 'rest of the world', predominantly young, low-income and living outside the West, falls into caricatured tropes.[30] Biased framings lead to biased datasets. Aid agencies have typically defined the next billion users as the 'virtuous poor'—users who are utility-driven and leapfrog their way out of their constrained contexts through innovative technology. Businesses have framed the global poor in recent years as natural 'entrepreneurs from below', ready to hack their way out of poverty and innovate in scarcity. Education policymakers view them as 'self-organized learners', who can teach themselves with EdTech tools and liberate themselves from the schooling institution.

These framings generate skewed datasets, which in turn legitimize these framings. The next billion users are very much like you and me—they seek self-actualization and social well-being through pleasure, play, romance, sociality and intimacy. These youth need the state and the market to step up and keep their end of the deal, to provide a security net and fair opportunities.

Disregarding our common humanity contributes to a failure-by-design approach and a graveyard of apps.

Male Universality

The gender dimension amplifies this already biased approach. Feminist activist Caroline Criado Perez argues how we are in a world where the default user is male, reinforcing the myth of 'male universality'. She points out how the lives of men have become standard representations for those of all humans, disfiguring our databases by a woman-shaped 'absent presence'.[31] She calls for breaking data silences by diversifying our media diet. Films, news, literature, the arts—all could benefit from how the rest of the world tells their stories.

This male-oriented culture feeds a gender data gap and influences how we design our systems, tools and services.[32] Those who are cisgender and male, able-bodied and middle-class serve as 'unmarked users', while the rest of us who do not fit the mould are marked as fringe, despite constituting the majority.[33] Crash test dummies correspond to the height, weight and body structure of an average man. Planners measure grocery aisles by the average height of a man, making the top shelves inaccessible to many women. Smartphones are too large for the palms of typical women. Office air conditioners are set to lower temperatures that favour men's instead of women's metabolic comfort, negatively impacting the latter's productivity. It does not take much to scratch the surface of any industry, urban space or product design to discover the gender bias underlying the logic of its design.

These biases gain a new lease on life as they seep into training data for algorithms. We have become alarmingly familiar with the impacts of these biased algorithms on access to welfare, jobs, education, healthcare and other essential services. Joy Buolamwini came to prominence with her article for *The New York Times* on racial and gender biases in facial-recognition software.[34] In an experiment at the MIT Media Lab, she used a white mask to show why facial recognition tools performed poorly in identifying Black women.

In recent years, female data scientists have doubled down as digital activists, taking on a disproportionate burden as anti-discrimination watchdogs when building AI systems. These commendable efforts, however, concentrate around racial and gender bias in the United States and Europe. What we need is an equivalent army of Global South advocates, activists and thought leaders to care for the rest of the world and ensure an inclusive internet. It matters not just whose stories we tell, but what stories we tell. History is written by the victors as the cliché goes, but what of the vast majority who yearn for their versions of the story, about their ancestral resilience, heroism, creativity and accomplishment? Sometimes it could be as humble as a story to humanize.

The Story of Poaching

Poaching strikes a chord in celebrities and public personalities. Prince William spoke about wildlife in his first speech as Prince of Wales in October 2022 to honour the legacy of his grandmother, Queen Elizabeth II. In line with the royal tradition, he championed the protection of endangered species in a rare emotional plea to the global community. He talked about the shocking loss of 70 per cent of Africa's elephant population in the last three decades, saying, 'At this rate, children born this year—like my daughter Charlotte—will see the last wild elephants and rhinos die before their 25th birthdays.'[35]

Prince William shares a crowded space with celebrities such as Jackie Chan and Leonardo DiCaprio as they campaign for tigers, polar bears and elephants. Ironically, COVID-19 improved the plights of less majestic wildlife, such as the bat, snake and pangolin.

COVID-19 also drew the world's attention to the lucrative wet market trade in Asia, with Wuhan acting as ground zero for this campaign. Just after the pandemic broke in 2020, China issued a temporary ban on wildlife trade and consumption, and kick-started the removal of pangolins from the official list of traditional medicine treatments.[36]

Nevertheless, conservationists face an uphill battle to curtail this illicit sector. Wildlife poaching is a highly profitable criminal enterprise, slightly less lucrative than narcotics, human trafficking and arms smuggling. A rhino horn costs more than a pound of gold or cocaine in the market.[37] The United Nations estimates that this trade is worth $7–23 billion a year.[38] This multibillion-dollar illegal industry comes with sophisticated high-tech logistics, global infrastructures and organizational processes, much like any successful multinational business. In 2012, the World Wildlife Fund (WWF) partnered with TRAFFIC, a global wildlife trade specialist, to combat the industry. Their strategy entails four pillars in the global wildlife trade chain: stop poaching, stop trafficking, stop buying, and inform and influence international policy.[39] Bottom-up forces like community involvement and protection works alongside top-down global policy initiatives such as banning sales of wildlife items and running public campaigns to shift consumption patterns. In recent years, the global supply chain of this illicit marketplace has become increasingly digitized and automated. Wildlife conservation organizations have reached out to the tech industry to combat this formidable problem—to fight fire with fire. The cliché 'crisis breeds opportunity' comes to life with Intel's TrailGuard AI project, a tech solution from the AI for Social Good initiative at Intel Corporation. The head of the lab, Anna Bethke, explains how by pairing tech with human decision-makers, we can solve some of our greatest challenges, including the illegal poaching of endangered animals.[40]

TrailGuard AI's cameras alert park rangers in real time when they identify suspected poachers entering conservation areas. These computational networks can process significant amounts of incoming security data at high speed and accuracy to detect and alert rangers at the national reserves. This project is one among many in the AI for Good movement, including the Microsoft-supported Elephant Listening Project, Google's Wildlife Insights AI and Alibaba's cloud computing for conservation.

Multiple tech companies, research institutes and non-profits are now utilizing machine learning in their conservation efforts. Drones, infrared cameras, satellite images, audio recordings and sensors are sourcing the data needed from the field to create more AI-enabled, targeted interventions to combat poaching, trafficking and sales. The question remains on whether AI lives up to its reputation as a saviour in the field of conservation.

Shoot to Kill

Figure 1.2. Double standards of empathetic design

I teach a class called 'AI and Social Design' in which I introduce students to current AI for Good projects. Students critically examine these projects and make their recommendations. In one session, we discussed Intel's TrailGuard AI project. I divided the class into small groups and, within minutes, the students were buzzing with ideas. After the break, I brought the teams together to share their thoughts.

All groups expressed concerns about the digital divide: how to sustain AI in low-bandwidth environments with limited connectivity where communities have basic digital literacy skills. Students questioned whether rangers could reach the scene in time to prevent the killing of endangered species. The class offered a radical solution—that Intel should enable their AI-based cameras in the African national reserve with drones that can shoot to kill poachers on the spot. This solution was efficient, albeit morally problematic. A perfect learning opportunity, I thought, and shifted context.

I then asked my students to picture a man ready to kill an endangered koala at the Amsterdam train station, which has a police drone with the capacity to kill. Should the drone go ahead and shoot the man to save the animal? Students responded with an immediate and vocal 'no'. We have a rule of law, they said, we cannot take a human life without due process, and not for an animal, even if endangered. I then asked them what made the two contexts different. Total silence.

This incident reveals that empathy is not an intrinsic quality—it is a learned process. It is a core part of intelligence that nurtures our ability to relate to another person within their frame of reference. We need to expand our frames of reference to break out of our default modes and include the world's majority in our emotional repository.

The Good Life

We know little of poachers' communities, their motivations and their aspirations. They are abstract, which makes them easier to demonize. In 2020, *Africa Geographic* embarked on a 'reality check' on poachers, to reveal the people behind the statistics—capturing poachers' voices to understand why they do what they do. The revelations were at times heartbreaking and unsurprising: the need to escape poverty, hope for respect, aspirations for their loved ones and opportunities for prosperous futures.[41] One poacher expressed his desire to send his firstborn child to school and give him 'the opportunity which I was denied as a child'. Another poacher shared his attraction to poaching as it promised a 'good life'—a nicer home, purchasing power, stability. He was tired of feeling like the poor one in the community. 'The way they [my peers] were behaving made me look like I am not man enough because I couldn't afford what they could. I was turned into a laughingstock in my community,' he exclaimed. Several others shared that they faced tough times due to the poor economy and had to resort to 'desperate measures'. The majority of the poachers interviewed perceived this line of work as low risk and high reward, given the easy access to these 'commodities' and their lucrative value. Poaching was a smart career choice given high unemployment rates, low legitimacy of the state to deliver decent job opportunities, and deep and persistent inequality.

Neuroscientists have confirmed that emotional and cognitive intelligence share several neural systems and produce deep interdependencies, which impact how we make decisions and behave.[42] In the age of AI, there is momentum among programmers to see if and how they can code empathy into machines. Computer scientists Michael Bennett and Yoshihiro Maruyama make the case that empathetic AI is explainable AI.[43] If we can explain how AI operates—its intentions, its logics and its patterns—we can address underlying issues of bias and redesign accordingly. To be explainable is to be accountable.

Users have lost patience with the 'black box' treatment of not knowing how their digital tools make decisions that affect their everyday lives. Companies are listening to pleas for data diversity.

For tech organizations, data diversity entails expanding the communicative repertoire through the data capture of facial expression, voice and gesture as proxies for empathy.[44] Emotions, however, have little meaning without their context. Diversifying our sources of learning remains a significant challenge.

Technology, by itself, cannot solve the confounding problems of unemployment, poverty, state corruption and social justice that poachers face. AI has a humbler task to fulfil. The premise of most AI-enabled anti-poaching tools is a basic one—intelligence gathering. It should assist institutions on the ground. These tools gather specific kinds of data from the field, tracking species and poachers for detection and deterrence. This, however, needs to connect with a whole spectrum of intelligences from the field that can ensure we build our systems to matter.

Intelligence on the Ground

Notions of humanity and humankind are trapped in a Western bubble. In 2020, I supervised a project on how conservationists use AI in Africa. A young media scholar, Natasha Rusch, took on the study of AI-enabled anti-poaching tools in different conservation areas in Africa. Our report was part of a larger initiative by Digital Asia Hub and Konrad-Adenauer-Stiftung (KAS) on AI trends in the Global South, launched at the onset of the pandemic.[45] Born and raised in Zimbabwe, Natasha spent time in Zambia and South Africa, as her family members were well connected to the conservation industry. Their networks helped open doors for us to connect with rangers to learn about their ground realities, including the use of AI-enabled tools. We engaged with NGOs, conservationists, technology researchers and companies working in different conservation sites in Africa, and asked them questions on motivations, missions, financial backing, strategies, community involvement, intelligence gathering tools, partnerships and the local politics that surround such projects. We wanted to know what the different beliefs about wildlife poaching were and why, the rangers' biggest challenges and strategies to count poaching, and how they used AI in their everyday tactics.

We found that everyone agreed that intelligence plays a vital role in the war against poaching. But what is critical data in this war of survival? Intel gathering is risky business. Rangers shared their stories of patrolling on foot in difficult terrain over massive areas. They face poachers who are armed militia; trained sniffer dogs are useful, but exorbitantly expensive. Going by foot in this game reserve is not optimal, as 'you have one in a million chance of bumping into poachers unless you have intelligence,' admits a ranger from the Malilangwe Trust, a Zimbabwean NGO.[46] These facts make for a good case to use AI-driven patrolling sensors. Patrolling, however, demands smart tactics. According to a ranger from Painted Dog Conservation, another Zimbabwean NGO, it pays to be unpredictable and keep poachers on their toes. They mix up their routines, change their hours of inspection, and shift their patrol pathways and times of operation. Typically, AI-enabled anti-poaching tools predict tactics of poachers, fed on past movements across these regions. Is machine learning capable of this dynamism with these ground-level tactics?

Intelligence Leakage

Tech companies are experimenting busily in this area. Predictive analytics of poaching activities uses machine learning and game theory. Microsoft's PAWS project, for example, provides real-time heat maps that detect activities so that the rangers can see which locations have higher risk levels than others and adjust their patrolling strategies. The real intelligence, however, lies in how rangers develop tactics to make themselves unpredictable. These human intermediaries transform data into intelligence as they leverage their deep experience, informant networks, shifts in animal movements and cultural rootedness.

Reliable informant networks are critical to such intermediaries. One of the biggest problems is 'intelligence leakage', of rangers becoming informants for poachers. A ranger from Malilangwe Trust explains: 'Like Kruger National Park, they brought in all this fancy equipment, tracking equipment. They've put in transmitters into rhino horns. And the very people who are looking after them, are using that to actually kill the rhinos.'[47]

Relationships of trust dictate the quality of intelligence. A ranger from the Panda Masuie Project (an NGO) explains that 90–95 per cent of poaching takes place on designated conservation property. Rangers and poachers are often from the same villages, sometimes related to one another. To go against your peer network is no easy feat. The informant networks can swing in either direction based on how you treat your rangers. On that end, the picture is dismal.

While tech companies work hard making AI for rangers, rangers are hard at work trying to get their basic needs met. A 2018 World Wildlife Fund report interviewed 7,110 public-sector patrol rangers at hundreds of sites across twenty-eight countries. The report revealed that 82 per cent of rangers had faced life-threatening situations while on duty.[48] Despite these high risks, rangers were inadequately armed and had limited access to vehicles and training to combat organized crime. Other grievances included insufficient boots, shelter and clean water supplies. Their pay was low and intermittent, work conditions were poor, and jobs came with limited or no health and disability insurance. Intelligence on poachers was not their key problem. Investing in the ranger's well-being would have paid more dividends.

The realities on the ground are that most rangers have rotary phones and limited connectivity and electricity. They resort to paper-based record-keeping. The more computationally dynamic the AI tools promise to be, the more investment hungry these tools will be as they demand quality digital and data infrastructures. The competition for funding is fierce. AI wins over salary hikes for rangers, better roads and vehicles, training and arming rangers, and investments in wildlife tourism and other alternative livelihoods. Even though AI wins the battle, it loses the war. Without addressing other related factors on the ground, a Painted Dog Conservation ranger points out why AI-enabled tools will not work. The ranger talks about how bad the roads are in the Kalahari area. To ensure the AI-based alerts translate to action, they need to invest in basic transport infrastructures:

> Imagine you receive a phone call, or a radio: there's been an incursion; there are poachers in the area and you need to pull from Main Camp; it is going to take four hours before you get to Makomo. And I don't know how many more hours to get to where the actual problem is.[49]

Moreover, cofounder of Rhino 911 Nico Jacobs suggests we need to get off the high horse on high-tech. AI should work in tandem with tech better suited to being on the ground:

> With a proper placing of radar, a helicopter with unlimited flying time, monitored fences ... Better electrified fences, thermal cameras on the fences, radar systems and a four-pilot team that works eight-hour rotating shifts. I'll decrease it [poaching] for you by 80 percent.[50]

If we step back, our questions get tougher—even existential. Is there a point to detecting poachers if rangers cannot get there on time? Why place all bets on new tech when old tech can still be powerful? How will users perceive AI tools if they go against their own interests? What value does AI have when detached from social intelligence on the ground? This is not tech nihilism or Luddism; it is social realism. Social solutions can help solve technological problems. Human and technical intermediaries come as a package. We need to zoom out to see the full complexity of humanity.

From Morality to Relationality

There is a reason why the tech industry leads the way on AI for Good projects. Tech leaders miss being the good guys, like in the early days of the internet. Today, Google's 'Don't Be Evil' motto reads like satire. This popular slogan disappeared with the branding of Google's parent company, Alphabet.[51] Similarly, the popular HBO television series *Silicon Valley*, which parodies the bro culture of California's Silicon Valley keynote on the next from 2014 to 2019, was no longer a laughing matter. The show's writer and producer, Clay Tarver, remarked, 'I've been told that, at some of the big companies, the P.R. departments have ordered their employees to stop saying "We're making the world a better place", specifically because we have made fun of that phrase so mercilessly.'[52]

The tech industry's credibility is at rock bottom. It has a proven record of generating new problems with old tech and promising to solve these problems with newer tech. This pattern guarantees a cycle of technocentrism. Think of the cringeworthy *aha!* moments among the geeks in Netflix's docudrama *The Social Dilemma* about how the tech industry controls, manipulates and polarizes its billions of users. Tech inventors of the hashtag, the scroll and other such digital addictions promise to save the world again as redemption. But tech creators need to stop trying to be messiahs in this game. They need to break away from this legacy of moral certainty, and binaries of good and bad.

Designers need to shift from morality-driven design that have grandiose visions of doing good. Instead, they should strive for relational-driven design that focuses on the relationships between people, contexts and policies. Like any relationship, it is the process that needs nurturing. Designers, programmers and funders can benefit from attending to, listening to, and reflecting on what users at the margins say and do about AI intervening in their social lives. Policymakers should bury the terms 'AI for Good' or 'AI for Evil'. Words shape design, so regulators can establish transparency by reframing AI as it relates to the values of diversity and representation, authenticity and trustworthiness, and explainability and responsibility. The media can stop debating whether tech can be an autonomous solution to the world's problems.

All tech is innately assistive, and investing in the human is investing in the machine. AI does not generate contextual intelligence; we—humans—do. We need to learn to live without our binary instincts. This pursuit demands a messy, but essential, relational solution.

2

Algorithms of Aspiration

A STREET VENDOR IN CHIANG MAI, THAILAND, CAUGHT THE internet's imagination with her creative way of attracting customers. The twenty-three-year-old figured out the attention metrics, wore nothing underneath her cardigan and got online to sell *Khanom Tokyo*, a popular street snack.[1] Posts on her Facebook page got tens of thousands of comments and shares, and the tactic boosted her sales. Morality aside, this was calculated creativity.

When the media speaks of the 'creative class', a concept pioneered by urban studies scholar Richard Florida almost two decades ago,[2] it portrays young, latte-sipping creative entrepreneurs and hipsters in global cities like New York and Amsterdam. Few outlets cover people at the margins in the Global South—second- and third-tier cities—hours away from the global metropolises where a majority of the world's young people reside. It is no wonder that there is cognitive dissonance in picturing future digital creatives being born out of poverty porn.

Global data trends say otherwise. A 2021 survey by Hootsuite and We Are Social, respected Silicon Valley creative agencies, suggests that 'the "next big trend" in digital won't emerge from a Western market'.[3] The creator economy of fifty million[4] strong is a powerful manifestation of how young people are taking control of their self-expression, learning, social networks and livelihoods. Youth in the Global South are particularly determined to be at the forefront in shaping the digital future, as they cannot afford to sit and wait for their freedoms and future work opportunities. A majority of young people live with crippling pressures to conform to patriarchal and authoritarian rules. Social norms dictate how they present themselves online, what they wear and share, who they tag, where they check in and what they say. These online behaviours can have harmful implications—from sexual harassment and trolling, to imprisonment and bodily violence.

Navigating such suffocating digital and social spaces demands extraordinary and everyday creativity. The youth at the margins are constantly at work to figure out how to game the public gaze, navigate censorship and use the algorithms to feed their self-expression while remaining outwardly compliant. Taboos are tempting, especially those that suppress natural teen desires such as sexuality and romance. The youth have little choice but to be online entrepreneurs, as governments fail to provide them with gainful employment. Creativity is not the prerogative of the privileged. The youth at the margins are a formidable force in driving digital creativity with their unmet aspirations. Everyday online creative acts can be found in their digital work and play, joy, informal learning, and efforts to build identities and solidarities, if we only stop and look.

Imitator Nations

Former Hewlett Packard CEO, Carly Fiorina, described China as an imitator nation in her interview with TIME in 2015:

> I have been doing business in China for decades, and I will tell you that yeah, the Chinese can take a test, but what they can't do is innovate ... They are not terribly imaginative. They're not entrepreneurial, they don't innovate, that is why they are stealing our intellectual property.[5]

Fast forward to 2023, where little, yet everything, has changed. While the US *Financial Times* covers Chinese corporate espionage, *China Daily* touts its nation's global innovation. The *Financial Times* reached out to the Federal Bureau of Investigation (FBI) to learn how much of a nemesis China has become to the United States by stealing trade secrets, pirating software and counterfeiting to the tune of $225–600 billion a year.[6] *China Daily* counters this framing by arguing that China is a hub of global innovation. The Chinese media reminds the world that in 2018, China entered the global innovation index rankings of the top twenty most innovative countries in the world.

China's list of achievements is impressive. It pioneered the world's first quantum-enabled satellite, the world's fastest supercomputer, the world's largest and fastest radio telescope, the world's first solar-powered expressway, the world's largest floating solar-power plant and the world's thinnest keyboard.[7] *China Daily* concludes confidently, 'No matter how much the country's creativity may differ from the West, China will lead the world as a global leader in science and technology by 2030.'[8]

We are witnessing a creativity turn in the Global South as indigenous innovators are looking at their people as creative assets, legitimate markets and content partners to help build their data products and services. Yet the imitator label remains a sticky factor that translates to few creative industries in the West tapping into these global shifts.

Figure 2.1. Creative caste system

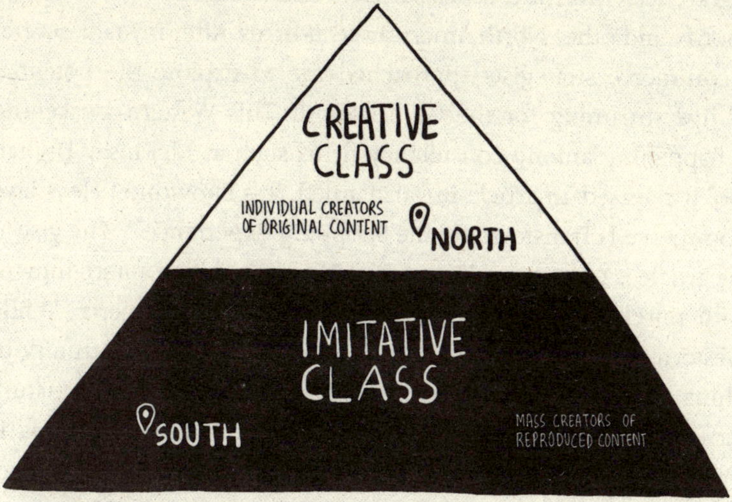

Creative Caste

Silicon Valley remains slow to recognize the geopolitical shifts in the creator economy. Nobel laureate economist Amartya Sen explains how the West suffers from a 'cultural captivity' syndrome, where a theory is born from the 'accidental correlation between cultural prejudice and social observation'.[9] Sen talks about the stickiness of how we imagine entire groups even when correlations die.

I call it the *creative caste system*, in which certain groups are believed to have an intrinsic ability to create original thought and content, while others do not. Groups are marked by the nation and the context they happened to be born into. People's gender, caste, race and economic status play a part in how they are perceived on the creative spectrum, and the poverty stigma, in particular, runs deep. It is still a challenge for Big Tech to view low-income users as consumers, let alone as creators.[10]

In 2022, I gave a keynote on the next billion creatives to UX researchers from the usual suspects like Meta, Google, Amazon, Spotify and other North American tech firms. After my talk, several e-commerce strategists approached me to explore the potential of live streaming for the Global South. This Western awakening is happening among consultancy firms such as McKinsey Digital, too. It released an article in 2021 titled, 'It's Showtime! How Live Commerce Is Transforming the Shopping Experience.'[11] The gist of the article is to convince Western businesses to blend entertainment with purchasing by using livestreaming for e-commerce. While Western tech is waking up to this supposedly novel opportunity, in China, this is old news. China's creator economy is now a mature ecosystem with a market size of $20 billion in 2021. China is already home to the world's biggest live commerce player, Taobao, with a market share of 35 per cent.[12] By 2020, almost 30 per cent of Chinese internet users shopped via livestreaming. Chinese apps realized years ago that content creators need to be able to monetize their talents and fanbases, or they would look elsewhere to set up home. US tech still grapples with interoperability and content monetization.

While Silicon Valley is still harping on blockchain and the nebulous Web 3.0, India has gone ahead and set up the Unified Payment Interface (UPI).[13] UPI is the Indian government's digital public infrastructure, allowing instant payments in peer-to-peer and customer-merchant networks. While crypto lords in the West battle each other over who will reign supreme, India has set up an easy-to-use, low-cost marketplace to help boost the creator economy and get the market working again. UPI has been so effective that Google has nudged the United States Department of the Treasury to consider it as a best practice while they design standards for open payment systems.[14] Tech business analysts Vivek Wadhwa, Ismail Amla and Alex Salkever argue that we should not be swept away by the 'cool' factor of new tech like blockchain. Instead, the real revolution of tech tools lies in the 'less flashy efforts' of open standards and interoperability, where India is way ahead of the game.[15]

More examples abound. Latin America is a driving force in e-sports and mobile game apps.[16] African entrepreneurs are pioneering 'data-lite' mobile apps to cater to their cost-conscious and mobile-first clientele.[17] As the world's eyes are fixated on Spotify to do right by musicians, Abu Dhabi has become a global hub for music-streaming apps, disrupting the global music industry.[18] Anghami, a legal music-streaming app, has become the first Arab tech startup to go public on New York's Nasdaq stock exchange. These Global South entrepreneurs have much in common. They are chronically underestimated, they must hit the ground running and deliver profits, and their price-conscious users demand added value for them to part with their money. Their only option is to think outside the box.

Out-Pirate the Pirates

Tech entrepreneurs in the Global South must innovate in environments where piracy is the norm, users come with constrained budgets and streaming is a challenge with unreliable connectivity. By 2028, 70 per cent of global subscriber growth for music-streaming services is estimated to take place in the Middle East, Latin America, Asia Pacific and Africa. Anghami's Lebanese cofounder, Elie Habib, explains, 'Our start was to out-pirate the pirates.'[19]

The Spotify business model does not work for Arab users because most Arabic music is independent and not backed by a label. Anghami pays more to artists and less to labels. The company launched as mobile-first for scaling across emerging markets and invested locally. They started by building relationships with labels and musicians in the Middle East before going global. Educating users on streaming, rather than finger-wagging on piracy, helped build a loyal base. The startup needed to make revenue from the get-go, given the unappetizing deals from venture capital funding.

Diversity and culture are strengths, not barriers. 'World music' is not some niche, but the norm. Habib believes that designers should cater to user aspirations and argues that music is intrinsically social. He remarks on how Spotify executives assume their users consume music alone when, in practice, users are deeply social and seek to connect with others through music.[20] Anghami found that most users shared music with others on WhatsApp and followed the playlists of those that matched their tastes to discover new music. His startup saw the potential in using machine learning to connect users with each other to build on this sharing culture.

While Spotify strategists remain unconvinced about the value of designing for mobile users in the Global South, Anghami leaders have built their business model around it, targeting the resource-constrained and those with precarious incomes. The startup takes this demographic into account by offering daily, weekly and monthly subscriptions, multiple pricing tiers, and access on low-end mobile devices and 2G networks. Given the high demand for audiovisual entertainment from the next billion users' demographic, and rising mobile uptake, Anghami allows music on par with audio. Habib explains the key difference between emerging and Western markets, 'It's not just about launching a service but providing the ability for people to try it, taste it and then eventually commit to it.'[21]

Western creative industries can flourish if they break out of their caste moulds. But public and private sector industries must first confront their traditional notions of who and what count as creative, and how to treat them right.

The ABCD Principle

In the last few years, several apps and platforms have emerged in the Global South with a business model that places the next billion users market at the centre. The classic case is Reliance Jio, India's major telecommunication platform, which made its data the cheapest in the world,[22] by catering to India's vast low-income populace. The NBU market was not an afterthought, but the core customer base. Reliance Jio designed its marketing plan by looking at the data usage of this user group. Marketers closely examined the kinds of content this segment consumed and produced and developed a marketing plan based on the patterns identified. It is called the ABCD principle—astrology, Bollywood, cricket and devotion—covering the four main areas young users pivot most of their data towards.[23]

Several Asian social media apps, like TikTok, Bigo and Kuaishou, grew exponentially across the Global South by targeting the underclass.[24] This demographic produced content creators overnight due to the app's ease of use and data efficiency on cheap phones. Most importantly, what these applications had in common was how their algorithms gave the global poor a fighting chance to be visible online. Their algorithms encouraged serendipitous encounters with complete strangers from across the world.

NBU-friendly apps assume an innate thirst for a diversity of stories and double down on their users' interest in fresh content. The apps enable users to rise through the social ranks, despite their socio-linguistic contexts, and the platforms offer seamless experiences for messaging, livestreaming, live video calling, and short video formats, allowing content to scale across borders.[25] As one Pakistani creator on Bigo, a free livestreaming app, states, 'I'm not even that famous, but when I go out on the street, people recognize me … People want to see new faces. They don't want to see the same old stars. They want to see someone real.'[26]

The digital economy in the Global South is in flux. There is a push for innovation, a pull from the underclass and slow movement from the state to rein in these disruptive forces through regulation. One thing is uncontestable—the Global South is riding the wave of the creator economy. This should not be surprising. If data is king, then these regions win the crown with their rising young populations going online. China and India alone constitute most of the users in the world, and both have yet to reach market saturation.[27] By 2030, young Africans are expected to constitute 42 per cent of global youth and are in line to significantly disrupt the creator economy.[28]

The economy of scale is fed by Global South users' aspiration and enthusiasm for change. Entrepreneurs have had to offer deep and added value to their customers in a highly competitive space—with few rules to protect them, little infrastructure to support them and limited numbers of users with disposable incomes. For decades, Global South economies have been predominantly informal and precarious, and, in many of these countries, formal employment is a rarity. Eight out of ten workers in Africa are informally employed, the highest share among all regions.[29]

Moreover, with less than 7 per cent of the world enjoying liberal democracy, according to the Democracy Index 2021,[30] users are forced to become more creative when online. They constantly pioneer creative tactics to circumvent AI-driven state machinery and online censorship regimes. These everyday hacks contribute to the ripening of the creator economy, contrary to some mythical nationalistic quality of innate innovators.

While the creator economy in the West is targeted at the urban elite, the rest of the world is opening up to the marginalized majority. While Silicon Valley views the global poor as high risk, Global South entrepreneurs are increasingly seeing them as opportunities long neglected within the monopolistic tech ecosystem. By building for and with the world's majority, you can scale up from the bottom.

Contested Creator Class

In 2022, Meta commissioned Richard Florida to revisit his concept of the creator class as it applies to the creator economy.[31] Florida, in his report, admits that we need to expand our vision on who these creators are. He stays close to his comfort zone of the United States and extends his paradigm to include the middle class—in response to a decade of criticism on his elitist approach to the creator economy. He admits that there is an 'immense opportunity to forge a "Creator Middle Class"'.[32] With the middle-income focus this time, the Southern creators at the margins are absent again. Florida spins the old yarn of how these creators are 'scientists, techies, artists, designers, entertainers and professional knowledge workers in fields like management, healthcare and law', calling them 'passionate hobbyists' who are not creating for money, fame or followers.

His model remains exclusive and out of touch. Who and what counts as creative needs an urgent rethinking by shedding the cultural baggage that demarcates the West from the rest.

Hacking Creativity

In 2022, a multinational design company hired me in order to gain insights on what digital creativity looks like in the Global South. They were looking to expand from their usual clientele of design professionals and other white-collar creative workers in the West. The company and I agreed to use India as a pilot study given its socio-cultural diversity, high digital uptake, and its vast, young creator economy. The goal for the pilot was to look at how creativity is defined, practiced, learned, shared, curated and perhaps even monetized among Indian youth. I was given free rein to design the project and built a ten-person team spread across different cities in India, that included design educators, videographers, digital ethnographers and media scholars.[33] I partnered with UX researcher Laura Herman to help shape a framework to approach this creator base and translate insights in ways that the design industry could optimize. Our framework looked at how digital creativity is shaped by access, identity, process, data, expression and learning.

Figure 2.2. Creativity framework

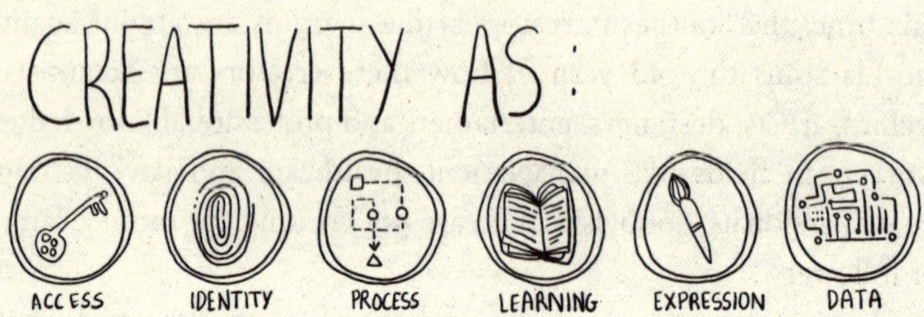

Our team investigated questions such as: What kinds of content do the youth aspire to create and why? Which design tools do they use? What was their creative process? How did they navigate issues of access, social and political surveillance, and other constraints, when creating? Did they identify as creators? How did they learn to use these design tools? What kinds of creator communities did they belong to?

To narrow our scope, we focused on Gen Z—those born between 1997 and 2012—and targeted two types of groups: one of learners and educators in informal and formal learning contexts, and the other of communicators, artists, designers, small-business owners, entrepreneurs, civic collectives and media organizations. Our designer-educator Siddhi Gupta led in-person workshops on hacking creativity for youths in different socioeconomic school settings. Next, we contracted product designer and influencer Trupti Shirodkar to host a masterclass on her Instagram and YouTube channels to harness insights from her followers.[34]

Our learners came from three distinct economic brackets: low-, mid- and upper-income schools. Diversity of the group was critical as we strived to balance gender, religion, caste, class and other intersecting identities. We spoke to almost eighty creators, including card makers, design students, influencers, digital activists, artisans, designers, ghost creators (intermediaries between creators, clients and companies), and small-business owners or 'micro entrepreneurs', who have their businesses entirely on platforms like Instagram (the cloud kitchen model, for instance).

A key insight from the pilot was that resource-constrained creators are the most enthusiastic, engaged and experimental with their creativity online. They break and make rules, and carve new pathways into the creator economy. Their creative journeys, however, face numerous barriers: access is limited, constrained, and comes at an economic and social price. Getting the Internet's attention is no small feat when you're a linguistic minority, or when your content doesn't resemble the neat Western aesthetics that pervade social media. Learning where to post, whom to tag, how to format and what kinds of voiceovers to do means that education must be gotten on the go. Far from being deterred, these creators were determined to make their mark in this digital creative space.

Access as a One-Stop Shop

Figure 2.3. NBU's criteria for app selection

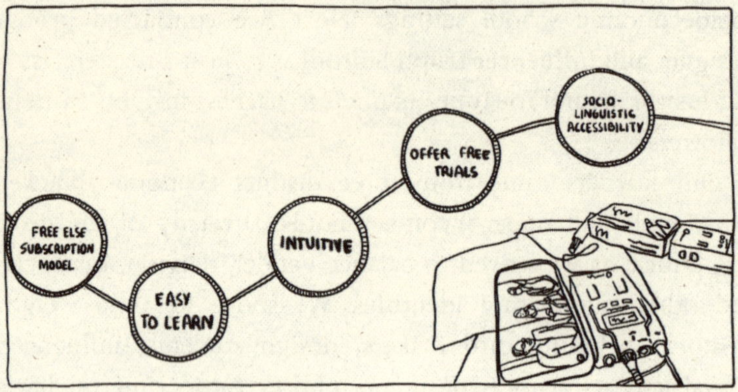

An eighteen-year-old design student works after her classes from 4 to 9 p.m. This leaves her with one hour for digital leisure every evening. In this limited time, she consumes and produces content. She does not want to spend a large amount of time learning how to use complicated design software. Instead, she opts for free, easy-to-use and quick photo-editing applications. Most low-income youth, including her, have multiple jobs to supplement their family income, and girls are required to help with household chores on top of these jobs. For youths like her, time is money. They consume and create content quickly, scheduling their time like CEOs on the run. Moreover, they are not just mobile-first creators, but mobile-only. The odds of them being able to afford a tablet or laptop in their lifetime are slim.

Their biggest demand is to have a one-stop shop of creative tools. For them, portability is key. They spend hours hunting for the right tools, apps, templates and software. Their criteria are that their suite of tools be free, intuitive, easy-to-use, mobile-based, and supportive of vernacular languages and designs. Their creative process is intrinsically social. They ask for tips on influencers to follow and YouTube clips to watch, and they trade ideas, run designs with peers, and basically chat on what is trending. Any tool that needs a lengthy tutorial is not meant for them, and due to their limited budgets, these youths don't appreciate complex applications designed to intimidate them.

A creative community of sorts mobilizes when in-application support fails, as instructions on apps are often in English. Youths crave a creative hub where all this action can happen in one place, a demand shared across socioeconomic groups. Users from the wealthier class must switch between multiple devices (phones and laptops) and types of software when editing audio-visual content. This makes the creative process tedious and ill-fitting with the demands of the contemporary creator economy. Free apps can be 'clunky' however, admits a media manager for a small NGO. He doesn't have his desired choice of templates or music overlays. Creating is a compromise when you don't have the money, as 'all the best stuff is like you have to pay for it', he says.[35]

Piracy is the primary solution to access issues. Resource-constrained creators prefer free applications that replicate licensed software, and not just for the cost savings. These creators need software that can run on smartphones with low memory capacity, using less data. In addition, they perceive expensive applications as tools for users with advanced skills.

Culture mediates access. Parental control is heavy among middle- and upper-class groups, though less so among low-income groups. Parents watch their kids like hawks and expect them to be utility-driven in their content creation. Creativity is suspect and confounding to these helicopter parents. Youths exercise creative tactics to gain access, free from parental surveillance. As one youth puts it, 'I use Instagram from my father's phone, but he doesn't know.'[36]

Profiles and devices are not necessarily shared due to resource constraints. They can be strategic choices for (in)visibility, especially when the content is politically contentious. While a super-app of creative convergence is a dream across groups, creators expressed their reservations. They wonder if an affordable creative hub is a pipe dream.

Niche Is Scalable

Creators hunt for online niches to go mainstream. Remixing niches can create virality. Combining poetry and food marketing, or Kerala pickles and adventure travel is the norm. As one aspirational influencer puts it, 'My poetry should reach more and more people. I want to join hands with products or companies, may it be food products or specific products.'[37]

Micro entrepreneurs such as 'Goodboy Kitchen' (@goodboy_kitchen on Instagram) combine the rising craze for special dog breeds among Indian elites with socially conscious marketing of plastic-free packaging—from Instagram to their doorstep. Cloud kitchen models promise Kerala pickles from their farm to the diaspora's table.

A niche can be simple, humble, grounded and yet gain global attention if it hits the right emotive nerve. Village fashion influencers like Neel Ranaut (@ranautneel on Instagram) from Tripura have gone viral and gained more followers than most of the top fashion brands in India. Ranaut combines stones, petals, leaves and flowers with adhesive tape and parades through their village. The cognitive dissonance of playing with gender identities, fashion trends and village stereotypes have hooked millions on their profile.

In recent years, many low-income youths have become influencers through body flips, lip syncing to popular songs and cool dance moves. Tanzanian village influencers Kili Paul (@kili_paul on Instagram) and his sister Neema Paul (@neemapaul155 on Instagram) have garnered millions of followers, from India in particular, as they move and groove while lip-syncing to popular Bollywood and Tollywood songs. In February 2022, the Indian High Commission honoured Kili Paul for winning 'millions of hearts' in India for his creative contribution.[38]

Small-time artists try their luck on Instagram by getting extremely specialized and contextual. Relatability boosts engagement. The local can transform into the universal by tapping into global aspirations for travel, food, entertainment, community and intimacy. Content creators need to carve their niches and stay with it, as one ghostwriter for a media organization explains:

> Whatever your niche, whatever be your category, on that basis you have to create content because people follow you from that point that this is a dancer, this is a food vlogger … On that you can't switch your niche. So, first decide your niche and then on that niche relevant content should be posted.[39]

Niches have limits. Micro entrepreneurs must balance scalability and selectivity due to limits in logistics. While their businesses may be entirely on Facebook, Instagram or WhatsApp, and promise unlimited growth, they still have limited offline production and delivery capacity. Artists have modest aspirations for follower counts since their price points are high and thereby affordable only for premium clientele. Civic activists may have specific causes that are tied to a mission or an event, which may not align with the medium of virality, as one NGO member shares:

> The entire social media game depends a lot on making Reels.[40] But as an organization, it also gets difficult to make

> Reels, because what are we supposed to do … the topics that we work on are so sensitive that we can't make it … you know, we can't always make certain things that goes along with the, you know, trending audios. But we can't dance to audio, right? We can't do the trends. So, we can't jump into every trend with the sensitive topics that we have. So, it gets very difficult to engage at times.[41]

Typically, though, aspirational influencers aim for masses of followers to attract advertisers, paid partnerships, affiliate marketing and ads on reels. They optimize many platforms, including Indian apps like ShareChat,[42] Moj,[43] Josh,[44] Nojoto[45] and Chingari[46], beyond the usual suspects, such as Facebook and Instagram.

Many content creators recognize that their content is niche. To survive in this game, creators need to be authentic to their context. Their cultural capital, where they come from—their village, or what they wear, or eat, or how they dance—can be turned into useful data if remixed with the right K-pop dance or Bollywood song. What was once considered 'primitive' is now prime for the algorithm.

Product Is Process

Creativity is collaborative. An aspirational influencer describes her creative process:

> When I have to create a masterpiece, a very visually appealing content, and I do not want to deal with learning how to use a difficult application, I choose to collaborate. In such a situation, I get to focus on the content part of the project, like the storyline/plot, and the other team member, who is an expert with an application, will manage the visuals.[47]

You can create a digital 'masterpiece' and yet share creative ownership. Young creators view themselves as creative 'geniuses' as well as collaborators. They don't view this as a contradiction; it is their creative reality. Aspirational influencers put tremendous energy into the creative process; staying trendy can be exhausting and putting their content out in the wild can be risky. It helps to have a creative community to transform concepts into content, test ideas in a safe space, and build relationships and buy-in with peers. Reciprocity is fuel to these creative fissures. Online groups play a vital role in reinforcing the collective creative identity of members. They help resource-constrained users access expensive applications and learn resources through designer-driven mentorship programs online, 'giveaway' awards in competitions and joint scholarships for licensed apps.

Marginalized creators find a cultural belonging for their creative expression. They seek out 'DIY Clubs', hashtag networks and curation communes. Followers also have a role to play in this creative collaboration. Aspirational influencers seek to co-create with their followers to keep the conversation going and deepen fan loyalty. Influencers and creators establish authenticity, trust and credibility by sharing their raw products with their audience. At the same time, creators' obsession with novelty gives way to remixing as the creative standard.

Remixing is at the core of self-expression and is well suited to digital cultures that are built through networks, cross-posting and collaborations. Creators view the process of building on other people's content as creative and critical in this trending culture. Youths are unapologetic and frank about their creative approach; as one of them explains, 'Remixing content, mashing things up and reviving old trends are very important because you do not always have something to say.'[48] Content without visibility is meaningless as one Gen Z youth puts it, 'What is the point of making [something] if I can't post [it]?'[49]

Culture of Credit

Attribution over ownership is the new ethical code of practice among creators. Rising creatives value giving credit to originators via online tags to acknowledge inspiration. It is flattering for many creators to be copied; as one remarks, 'I feel good that someone is copying. I feel that I must have at least done something good to warrant that.'[50] However, it is inexcusable to not give due credit.

This is felt most acutely among rural and semi-urban creators. They are deeply tuned into the copyright regime as they have been at the receiving end of intellectual and creative theft from mass cultural industries for decades. They want content protection, like watermarks and fair remuneration systems. However, in the face of these challenges, one rural influencer defiantly remarks:

> Today they [other content creators] might copy and take one or two creations of mine, [but,] finally they will have to do something on their own. How long can they be dependent on me? If they take one or two palms full of water from an ocean, what difference will it make?[51]

Increasingly, creators are demanding that attributions be factored into the design of the platforms. With rising protests by influential creators of colour, TikTok's director of creator community, Kudzi Chikumbu, in a 2022 press statement announced the new 'culture of credit'.[52] The goal is to introduce features that will make the algorithm more equitable for underrepresented creators. Chikumbu promised creators that they would finally have crediting tools like the ability to directly tag or mention in the video inspirations behind their content. TikTok has added user prompts to nudge creators to give credit, educating users on the importance of crediting via features like pop-ups.[53] TikTok also debuted an 'originators series' that highlights trendsetting creators on their site every month to build awareness of creative attribution for a global creator community.[54] While it is a step in the right direction, this does not fundamentally align with redistribution or profit sharing between creators and platforms. Moreover, attribution is still talked about in terms of individual creators online, with little attention to the offline social networks of creators that enable these creative products and processes.

Creativity has long been a collective process.[55] This is especially so among artisans in the Global South. Our algorithmic cultures, however, are biased towards computing individual people's products and processes. For instance, in Bangladesh, women make up 60 per cent of the labour force in the creative industries but barely eke out a living from their creative work.[56] Prominent heritage crafts like *kantha* are made by rural women who gather in a household and repurpose worn-out clothes with intricate embroidery. Ownership, here, is the village where it is produced.

The pandemic pushed many of these artisans to sell their creations online, inspiring several social entrepreneurs, like Shimmy and Bengal Muslin, to step up and equip these women to leverage digital resources and boost their living. Many women artisans have had to rely on male family members who had better digital skills and access to mobile phones. These new intermediaries have resulted in further cuts from their already meagre livelihoods.

There is a double standard on platforms like Etsy, which claims to be a digital space for 'millions of people selling the things they love'. In practice, its typical creators are middle-class, white women, who do artisanal work as a hobby. Their craft is recognized as personal, handmade and authentic. However, if the kantha women's groups came on Etsy, they would be viewed as commercial and mass produced. In conventional design hierarchy, individual and unique creations trump mass products and collaborative processes. Templates are out to disrupt such hierarchies.[57]

Template Is King, Vernacular Is Queen

Figure 2.4. Desi vs foreign aesthetic

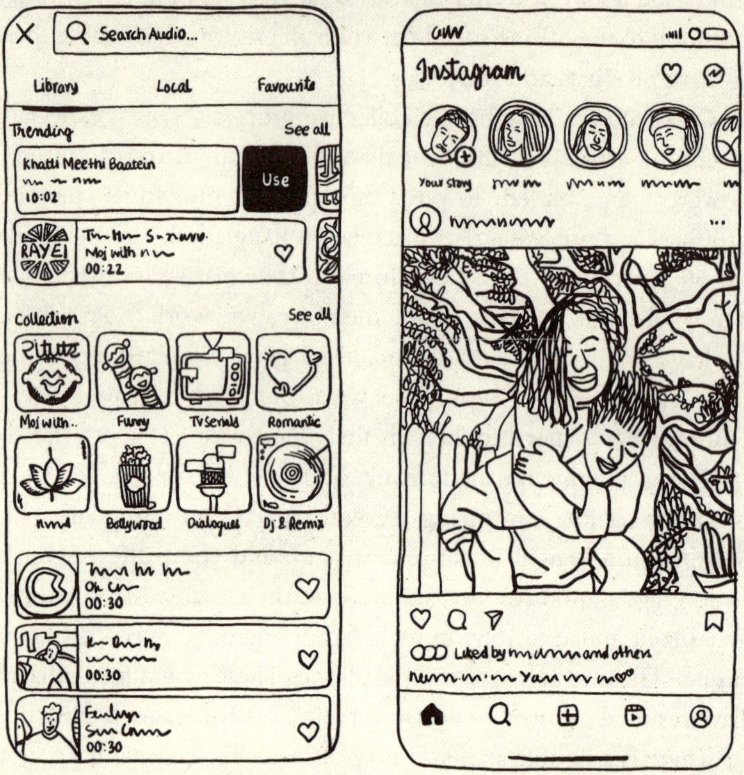

Western designers hate templates.[58] Many see templates as the beginning of the end of creativity. Then came Canva, cementing the hate. Canva is an Australian multinational graphic design platform that is used to create social media graphics and presentations, offering users with easy-to-use, ready-made templates that they can customize to produce professional designs. Founded by Melanie Perkins in Perth, Australia, in 2013, this template-driven design platform was valued at $40 billion in 2022. With a monthly user base of 75 million, it is considered one of the most influential design apps worldwide. This mass production of creativity has many designers in Silicon Valley appalled.

Drew Clemente, a systems administrator engineer, captures this angst in what he calls a 'dangerous trend' of bad design. In a rant titled 'Why Do Designers Hate Canva?', he speaks on behalf of his cabal by accusing Canva of producing designs that are crude, chaotic, unpleasant, garish, tacky, amateurish and lacking in sophistication.[59] He laments that their business model perpetuates a new aesthetic lacking in 'basic design principles'. He detests Canva's premise that 'design is easy', and sees it as a direct insult to professional designers whose careers are threatened by this new worldview. He is convinced that Canva has made 'the world a worse place' by making it harder for people to appreciate good design when they see it.

Against this passionate sentiment, the young content creators we interviewed in India saw things differently. They swear by Canva. They do not believe in suffering through a rite of passage to reach a 'designerly' level. They want to hit the ground running. They don't think design is easy; they *expect* design to be easy. Amateur aesthetic overrides sophisticated designs in allowing creators to respond to trends in a timely manner and showcase their authenticity. Design principles, for them, take a back seat to what the trends demand. These youths have no intention of pretending to be professionals. Their amateur status sometimes gives them more credibility among their followers. What may be garish, tacky and unpleasant to designers schooled in Western aesthetics, may just work in the Global South, given the right mood, theme and context. Creators place a premium on emotion as a driver for creative expression.

Across income groups, creators prefer using templates for three reasons: templates enabled the foundational work of content creation, allow for aesthetic consistency and make it easier to cross-share. Low-income creators are avid fans of these ready-made formats and features. They insist that contrary to the idea that templates reduce creativity, using templates frees them from technical functions and carves out more time for them to be creative. On the one hand, creators want to be original, but on the other, they want to fit in, and templates allow them to do both. Customizing templates then helps creators build ownership and originality. As one creator states, 'If there is a template which has ABC elements, I try to remove 2–3 elements and then I edit … and make it my own and then I post it.'[60] Some creators interpreted 'templates' to mean distinctive styles of content posted on their accounts.

Every bit of a creator's decision-making personalizes their templates. Their choice of captions, tags, filters, hashtags, colour schemes, post timing and occasions contributes to making the template their own. Creators in NGOs recognize that some designs can be off-putting to their target audience and very arty or out of reach. There is a class dimension here; creators in NGOs are often middle-class, while their audiences may be from a lower economic segment. 'We try to make it as bright and nice and beautiful as possible in a way that doesn't isolate or alienate the people.'[61]

Desi versus the Instagram Aesthetic

With twenty-two official languages and almost 20,000 dialects in India, local or vernacular design is on the rise, catering to a variety of socio-linguistic needs.[62] But vernacular design is not confined to linguistic needs; it also valorizes diverse cultural aesthetics. 'Desi' (a colloquial term for Indian) aesthetic is becoming increasingly popular among civic collectives, artists and a broader swath of middle-class youth in India. Young creators feel cultural pride in heritage designs and colourful palettes. They recognize that Western minimalism is just one aesthetic and not the idealized taste marketed in design schools over generations. However, creators complain that some of these indigenous apps are too busy, dense and unintuitive in their interfacing.

Aesthetics are not abstract principles. They are tools for mobility. The 'Instagram aesthetic' is an aspirational aesthetic, especially for creators from rural and semi-urban areas. It gives their expression a 'foreign vibe' with its minimalism and use of whites and soft palettes. Marginalized creators see themselves as global citizens and make social media their digital neighbourhood. This doesn't negate their cultural capital. Creators generate desi templates for the numerous festivals, rituals and events celebrated in India. Rural creators like their bling as it captures the festive mood. However, if the mood is more thoughtful, they adopt a more sober aesthetic. They try out drafts on indigenous short video apps like Moj, Josh and ShareChat, before going 'public' on Instagram.

Global South creators are aware of the creative tax they pay when using vernacular-based design. Western platforms' grammar of virality does not easily allow for a diversity of languages, cultural slang, visual representation of emojis and themes. The obstacles of font options, formats and language syntaxes push them to strive for a purely visual aesthetic. At the end of the day, however, engagement metrics rule the roost and dictate aesthetic choice. This demands constant learning and applying if creators are to stay relevant, connected and visible.

Jugaad Learning

I have previously written about how the global poor have been celebrated for 'hacking poverty' through jugaad, an Indian term that means 'to improvise, particularly as a response to scarce resources and rigid social systems'.[63] The term has entered the Oxford dictionary that emphasizes its innovative angle—'the use of skill and imagination to find an easy solution to a problem or to fix or make something using cheap, basic items'.[64]

Jugaad as a creative practice is intrinsically cross-cultural. Chinese people call it *shan-zai*; Brazilians call it *gambiarra* or *bacalhau* and Americans call it 'hacking' or do-it-yourself (DIY). Jugaad strives to be more time-, labour, and cost-efficient. As a product, it implies an affordable replica applied in different contexts, or an alteration to and even an improvement, on an existing feature, function or entire design.

Jugaad through informal learning is increasingly taking on new and creative forms. COVID-19 disrupted formal public education and legitimized DIY learning among the youth.[65] Creators' choices online have expanded significantly, helping them build their own 'curricula' of mentorship, knowledge, and networks. Many low-income youths perceive lengthy online tutorials and formal curricula as out of touch, slow, inefficient, tedious, and unnecessarily complex, viewing these traditional methods as'passive learning'. To produce as you consume is the active way. Creators learn through YouTube videos and Instagram reels, with trial and error leading the way, as one aspiring designer explains:

> I think training is not the right word, maybe practice like if I have ... For example the first time I did embossing, I learnt it from a video that I saw online, the first time I did it, it was a bit weird, and the embossing was falling off from places.

> Once I did that technique five or ten times, with practice I got the hang of it.[66]

Another creator shares her challenge of trying to complete an online Domestika course[67], but dropping out due to lack of time. She then joined creative communities online and started to fiddle around with different interfaces, to see what sticks and what doesn't. 'If it [online feature] works then I make it like a regular thing in my job. If X feature is working well and the output is nice, I'll continue using it. If it doesn't, then I scrap it.'[68]

Self-Taught Is Cool Yet Cumbersome

Self-teaching has become 'cool' across different groups. Creators gain respect from their peers when they learn by doing. Creators' DIY education can cover how to use software tools, set up a profile, and employ different tactics to increase reach and engagement. Experimentation unpacks the metrics that dictate their daily curricula. Since digital creativity demands nonstop content production and analysis, creators put a lot of pressure on themselves to be consistent in sharing content. Some creators record videos of their work to analyse later and gauge how far they have come: Why does one post gain attention while another post is ignored? What format works well for which type of content? What kinds of popular content forms should be remixed to optimize engagement? An aspirational influencer tries to figure out, 'If I use a song that has only 7,000 people using it, that might do well because that will show up on people's pictures. And then it depends if I show my face. I get more views when I show my face.'[69]

Many creators strive for a formula to get the engagement right. A 'successful' creation depends on the right mix of likes, comments, shares and views that the posts generate. Some creators get specific, saying, 'The engagement rate should be about more than 4 per cent. Means above 4 per cent it is best, means your content is working.'[70] Some bet on the post's timing to get it right. They have learned the optimal and active follower time the hard way. As one influencer explains, 'If I posted in morning, I get between up to 10, I guess 10,000 views ... afterwards, it goes up to 50 to 40,000 on Moj.'[71]

Learning isn't a lonely activity. Creators need communities to learn from and with to feel protected and encouraged and to figure things out together. Many realize that being a creator is a 'relatively lonely career'. They seek mentors to review their portfolios online—artists want others to help them price their art; designers want tips on trends, tools and templates for inspiration; NGOs dealing with vulnerable communities want advice on how to protect their clientele through anonymous messaging, strategic blocking, and other tactics. Women creators seek solidarity as they are disproportionately harassed online. They tap into their forums to tackle issues such as sexual threats, as they face privacy breaches more frequently than their male counterparts. Creators receive a barrage of messages on social media platforms and rarely reply, constantly trying to learn how to manage their communication. They are also inundated with live show requests. Engaging with these kinds of varied requests can be tricky, especially for women creators, as it opens a pathway to inappropriate advances and comments. Creators are, therefore, eager to learn when to use video options and when to close tagging.

At the same time, jugaad learning doesn't negate formal education. It comes down to desirable networks, time with mentors, and quality feedback. Creators talk of the struggle to get this support online; as one aspiring designer explains, 'Learning a tool is a DIY thing, but your designs need a qualitative evaluation. You need someone to tell you what will work and what won't.'[72]

Creators learn quickly that you need to perform or perish. They learn the hard way to gamify themselves, their emotions, their cultures and their bodies, sometimes to their detriment. Their pedagogic toolkit is stacked with tips on navigating social harassment, especially if they are a part of the marginalized majority. Their communities have taught them ways to build solidarity, support and resilience against platforms for fairer remuneration. Despite it all, they are optimistic about their digital futures. Their creativity is a mirror of their aspirations.

Next Billion Creatives

India is just one case among many Global South creator economies where we see creativity play out online every day. While on average 20 per cent of internet users worldwide follow an influencer, in Brazil, the figure is double that. A consumer survey by Statista, a data company, shows that two-fifths of Brazilians say they bought a product because of an influencer, the highest share among the fifty-six countries surveyed.[73] Nielsen, a market-research firm, estimates that Brazil has 5,00,000 potential influencers on social media, more than anywhere else in the world.[74]

In fact, these companies have discovered that the next billion creatives drive the creator economy. Creators from resource-constrained contexts are taking the lead, being self-driven and self-taught, given their limited options. The digital has created a niche that can help these creators scale. Southern creators pioneer clever shortcuts to enable their creative processes and learn on the fly due to limited leisure time. They are unapologetic about their economic status and, in fact, take pride in producing more with less. Their expressions find a virtual home and resonate with a global creative community, despite the dominance of the Western aesthetic on social apps. While their privacy matters, their visibility matters more. They are acutely aware of the fierceness of the attention economy and often use their data as currency. They are hungry for followers, likes and comments, and keenly study their metrics to remix and reshare their posts to get it right.

However, many refuse to call themselves 'influencers' as they reserve this title for those able to monetize their content and earn a livelihood from it. A few detest the traditional view of an 'influencer' who tries to peddle brands and get people to 'buy stuff'.

Most did not expect to be an influencer in the conventional sense of being able to attract hundreds of thousands of followers. Creators want to get rich and famous from content monetization, but are realistic about their prospects. Their aspirations are to nurture a community beyond what they are born into, to have fun, gain status and, most importantly, make a mark 'out there', through their digital expression. These creative climbers are aspirational influencers. They live by a code of digital creative practice, and want to be recognized and respected by an online public. They believe that when you are true to yourself while creating content, you become relatable to others. They want to be seen and heard. It's their entry point to a better future.

3
Digital Desire

'RAISE YOUR HAND IF YOU DO NOT WATCH PORN.'

Audiences hesitate, then laugh uncomfortably. Almost nobody raises their hands. I bring up this question in my talk circuits in Hamburg, Panama, Copenhagen and Amsterdam when speaking about future digital trends. This provocation evokes the same response across audiences—digital-marketing specialists, entrepreneurs, policymakers, investors and programmers.

At one telecom conference, a manager comes up to me after the event and tells me that I talked about the 'elephant in the room'. He says his Latin American team deliberately ignores customer metrics on mobile data usage for porn. This way they don't need to disclose these trends as doing so may cause panic among state officials. Online porn keeps their businesses going and enables mobile uptake among digital newbies—the tech industry's best kept open secret.

Figure 3.1. Porn as a gateway to the internet

A health and development magazine asks me to edit my op-ed and delete a section on sexuality in healthcare in developing countries. In the editorial, I had written about how healthcare has been historically geared towards population control and family planning instead of towards normalizing sex and sexuality as essential parts of human experience. Youths want to discover themselves and their desires, and these metrics can reveal honest patterns of young people's intimate needs and wants, especially in Global South countries where sex is taboo. The editor agrees but says she is concerned about their public health readership, which is focused on family-planning interventions and has more conservative values. Sex is like Voldemort: that which cannot be named.

Despite society's moral panic, porn is and always has been the gateway to the internet. Thirty-five per cent of all internet downloads are related to porn, and women make up one-third of porn-viewing audiences.[1] Few public institutions provide content that speaks to youths coming of age, especially about sex. The youth have few people or places to turn to for information or reassurance about their bodies, sexualities and sexual insecurities. It is no wonder that porn has become an educator for teens across the Global South.

We need to shift the war on vulgarity mindset by making peace with sex. Policymakers and tech entrepreneurs should take a pleasure-positive approach to sex and sexuality on the internet that considers normalizing sex and recognizing its stealth data politics.

Indecent Proposal

Governments, especially in the Global South, are proactive about moral controls. Ministries crack down on 'indecent' content. Officials instruct their information technology (IT) divisions to weaponize content moderators, digital vigilantes and machine learning technologies to act as the moral police for the digital age. The definition of porn has expanded in patriarchal states. For instance, influencers in Iran face long jail sentences for promoting 'vulgar' content that spans workout videos, dancing and fashion.[2] China hunts down livestreamers who post homosexual content, which their government perceives as 'obscene'.[3] Kissing in public is taboo in India and can result in obscenity charges.[4] Indonesia blocks thousands of porn sites and other websites deemed immoral. This is part of its Internet Sehat (Healthy Internet) programme, which strives to reduce 'prohibited information', including 'blasphemy' and content that 'creates community anxiety' and 'disturbs public order'.[5] According to the 2022 Internet Censorship report comparing global internet restrictions, 82 per cent of Asian countries restrict online pornography, and twenty-seven of them have instituted full bans.[6] These restrictions include bans on torrents, pornography, social media, virtual private networks (VPNs) and messaging/VoIP apps.

Similarly, the Gulf region exercises moral control on the internet by blocking adult sites. Even browsing these sites is a serious, punishable offense. Yet six Muslim states rank among the ten top countries that watch the most porn.[7] These figures are underestimated as sex-related content has pervaded social media and is hard to track. Encrypted spaces are thriving with the sharing of intimate, and often nonconsensual, sexual content. Whether we are trying to shop, pay our bills, chat with a friend, or look up the news, sex has a way of getting our attention online. Despite these moral controls, the youth make headway on the internet in producing and consuming content on love, sex and relationships. And the youth in the Global South are much like teens worldwide, for whom sex and sexuality are fundamental pathways to self-development.

Metrics of Intimacy

Facebook campaigns on sex and pleasure in Egypt, Kenya and Mexico appear visibly dead with almost no likes, comments or shares. Yet activists are encouraged by their millions of loyal stealth followers, tracked using back-end metrics. The Facebook page 'Love Matters Naija', a Nigerian non-profit initiative on 'love, sex & everything in between' is a case in point. Data on audience engagement allows this agency to shape its content on sexuality to feed this silent revolution. Engagement metrics and intimacy proxies are the landscape of digital desire in the Global South. Those who do post on Love Matters Naija, however, are frank, open, unfiltered, and can stir up heated debate on sex:

> Nipples are not an erogenous zone for everybody. It is not everybody that gets turned on from nipple stimulation. It is

> better not to assume that sucking the nipple will just turn anybody on. There are some people that can reach cloud 12 if their nipples are sucked. There are others that even if you suck from now till the end of this year, nothing will stand or get wet. Which one of these two types of people are you?[8]

One man responds, 'Mere seating close to my partner turns me on', while another remarks, 'A man that has money and know how to segz. Anywhere he touches on lady's body, that lady must turn on by force.' A woman complains about men commenting on this post instead of women: 'My question is why is there no lady answering the question bcoz I think the question is for us.'

Set up in 2018, this page gained 4,00,000 fans on Facebook within eighteen months. The Arabic page is even more popular, with 2.5 million followers in 2022, and some of their videos watched over seventy-two million times. Compared to the Nigerian page, the Arabic page has even fewer comments and likes.

In June 2022, I was invited to celebrate the seventy-fifth anniversary of RNW Media, an international digital media organization, based in Haarlem in the Netherlands. It was RNW that established the Love Matters programme in India in 2011 and then scaled it across Asia, Africa and the Arab world through a social franchise model. Subsidized initially by the Dutch Ministry of Foreign Affairs, the campaign grew in numbers, funding and collaborations, and continues to help millions of young people have open and respectful conversations about love, sex, and relationships that are otherwise censored or considered taboo.

For the anniversary, local leaders of the Love Matters initiatives in Burundi, Lebanon and the Arab world were flown in to share their experiences of growing this movement online. Sandy Abdelmessih, project manager for Love Matters Arabic, explained how her team updates their strategies daily by looking closely at the metrics, which often don't match the comments section. They have found encouraging signs of high youth motivation for their content. As Abdelmessih remarks:

> There is an assumption always that you cannot speak about or put content about masturbation online because people would feel like this is so inappropriate and you're encouraging people to do behaviours that are prohibited by a religion and that does not fit our culture. But surprisingly, the pieces about masturbation are always on the top visited pages on our website and we receive around 18 million page views per year.[9]

While millions visit and follow their platforms, she states that Egyptian officials argue against circulating 'these kinds of topics' in the public, pointing out that people will learn about sex after marriage. She concludes wryly, 'Obviously all Egyptians do not masturbate.'

Risky Content

My FemLab team interviewed several NGOs in India between March and June 2022 to understand how they create digital content to engage their audiences on matters regarding sexuality and found similar challenges. One non-profit expressed frustration about how moralistic some people can get online, which leaves little room for an open conversation about sex.

> There needs to be a place for people to put out their personal narratives without them being held up to such perfect political standards ... I know people have got little pissed at any content about ... like the topic of porn really kind of angers people. Because everyone immediately imagines

> about the worst-case scenario of possibilities, but our position has always been okay it is a part of people's lives, we can't just ignore it and directly add to the shame that is already there.[10]

We found that platforms like Facebook and Instagram are more conservative when it comes to sex. Their default strategy for 'risky' content is to censor, ban or make it invisible. An activist shares her experience:

> I mean sometimes [digital platforms like Facebook] don't even allow there to be drawings of nude bodies. And even like the kind of posts that we can promote ... like a video on abortion is just impossible ...we had to change the text, not mention sex or abortion directly in the caption in order to promote a video which also almost defeats the purpose, because it gets classified as solicitation.[11]

Mahwish Gul, a development management consultant, explains why we need an app just to talk about menstruation in Pakistan, where two-thirds of teenage girls are unaware of menstruation at the time of their first period. This knowledge deficit creates anxiety, guilt and distress, as expressed by a thirteen-year-old girl who says, 'I thought this only happened to me. I thought my mother would be angry to know this. I thought she would say it happened because I did something to myself.'[12]

Gul collaborated with the East West Center in Hawai'i and the Center for Communication Programs in Pakistan to develop the #HelpSaira app. This tool uses storytelling to get teenage girls, parents and other community members to engage with common prejudices against menstruation and offer model responses to combat notions of wrongdoing. Alongside many sex education organizations and activists, Gul faced an uphill battle to bring topics around sex, sexuality and sexual preferences into the open. Patriarchal institutions saw these efforts as 'foreign imports' that challenged their culture and traditions. Yet metrics gathered from user responses to the app have the potential to counter these falsities and push the state to initiate and shape public awareness campaigns around these taboo subjects and perhaps establish sex education programmes in schools.

Toxic Taboos Online

Metrics-based activism is not a standalone technique of outreach. NGOs have found working with artists, influencers and entertainers and organizing in-person festivals and workshops to be effective in getting their campaigns out. Mediators navigate the balance between nudging people to keep an open mind while not imposing their own views. Organic approaches with targeted content using metrics such as gender, age, location, engagements and popular themes based on page views help reach the most vulnerable groups.

According to a 2020 study from the Pew Research Center on global laws and norms around same-sex marriage and LGBTQIA+ rights, public opinion remains divided according to country, region, and economic development.[13] While significant progress has been made across regions, including South Africa, India, and South Korea, toward acceptance of homosexuality, factors such as religion continue to present obstacles to acceptance.

Taboos of touch are complicated by cultural factors like caste. The morality around 'untouchability' is further structured by the caste system in India. As globalization scholar Arjun Appadurai explains, 'We live in a world where the senses have both expanded and shrunk' based on our caste positioning.[14] In this immersive digital world, while digital tools can allow for cross-caste contact and communication, they also enable targeted contact of Dalits, the so-called 'untouchables', through invasive biometrics of fingerprinting, tracking, carding and other surveillance. In such cases, Appadurai argues, 'Social warmth is not only not welcome, it is dangerous, since it might well promote other sorts of intimacy which endanger the haptic order of caste.'[15]

Sex taboos, institutional silence, and cultural shame have pushed young people to seek alternative spaces to learn about sex and sexuality. Many youths view porn sites as places that encourage society to perceive women as sex objects and set unrealistic standards and expectations. Yet they still seek them out as a refuge for sex education, due to the lack of legitimate alternatives.[16]

Pornhub Education

Informal learning has a long history of filling the gaps missed by formal education. Channels like YouTube, TikTok and Instagram have become the go-to portals to learn how to fix a pipe, renovate a roof, braid hair, do makeup, improve your pronunciation and fill out forms. British journalist Chris Stokel-Walker describes YouTube as having a 'do-it-yourself ethic; it is punk TV for the 21st century'.[17] Today we all expect to learn from YouTube. What is less expected is seeing youth flock to Pornhub to learn about not just sex but even calculus.

One of Pornhub's most popular global trends in 2021 was the 'How to …' search.[18] The youth turned to this site to learn about sexual positions, anatomy and pleasure points, searching 'how to touch a woman', 'how to use a condom', and other burning questions around sex. Some even became sex tutors and hoped to monetize their content. In 2021, Pornhub witnessed instructional searches increase by 244 per cent, with more than 2,00,000 videos tagged or titled as 'how to' videos. Sex coach Stacy Friedman claims that since the pandemic, there has been uptake from people keen to learn how to enjoy their sexuality.[19] Pornhub took notice of this trend and set up a 'Real Sex Education' portal.[20]

Pallavi Barnwal, a sex coach for Indians, explains how difficult it is for Indian couples to talk about sex and how that often results in damaging their relationships.[21] This is despite India being one of the largest porn consumer nations worldwide, ranked third after the United States and Britain. Barnwal learned the hard way how important it is to speak about issues around sex, having been brought up in a conservative household. She explains how much of our anxiety around sex is not just in the act itself, but in the full ritual around it: 'I get dozens of messages from people asking me what to do on a wedding night—not just physically, but how to appear not too shy and not too experienced.'[22]

Pornhub is not the only one listening. #LearnOnTikTok is one of the most popular hashtags on TikTok, and according to Pentos, an analytics company, videos on this channel have been viewed more than 412 billion times.[23] TikTok has become proactive in its sex education agenda with 'sex positive' channels and creator spotlights for these accounts. The channel 'Asian sex education' has almost 400 million views; on it, 'relationship experts' address topics such as how to use a tampon, how to come out to your parents, kissing during sex, long distance tips for sexting, issues with body hair, sex talk, dealing with anxiety, and even learning math.

You may luck out if you are willing and able to get in front of the camera and speak plainly and without condescension on what you are passionate about. Some content creators have realized that the best way to reach the youth is to go where they hang out digitally. When classrooms and tutorial centres shut down during the COVID-19 pandemic, many tutors were on the brink of despair. Taiwanese math tutor Shun-Wei Chang, who goes by Changhsu on his channel (@changshumath), contemplated suicide because of his mounting debt.[24] Instead he decided to shoot videos teaching calculus. He quickly realized the competition online, when it came to education videos, was fierce: EdTech had exploded. He set up shop instead on Pornhub with his channel's slogan, 'Play Hard, Study Hard!' It worked. His videos have racked up almost two million views, and students now purchase his course packages on his website.

Over the years, content creators have learned that once you have gained an audience's attention and trust, it is easier to become a guru on other subjects. Amateur creators school themselves on various trends in politics, fashion and cryptocurrencies to build up to super-influencer status. They view the world as their digital oyster. Take the case of the Peruvian fashion influencer Salan, (@salandela on Instagram), who shares her opinions on fashion, politics, abortion, sex, racism, women's rights, and other socially contentious issues. In her 2021 interview with the Peruvian journalist Jimena Ledgard for *Rest of World*, she remarks, 'I am rather lucky that brands are all about "disruptive" influencers now. They have embraced my political positions, even if they are controversial.'[25]

It is a sink-or-swim predicament for influencers, since followers punish them for their silence by unfollowing them. Gen Z Peruvians are at a pivotal point of political awakening, given Peru's social upheaval and numerous corruption scandals. The youth today demand that influencers align with them and take a stand on their issues. Content creators, therefore, walk a precarious line between virality and staying hidden from state censors. Being in the wrong place at the wrong time can become a career killer. Sometimes, when an influencer takes a stand, they might end up being 'invited for tea' by the state. Talking about sex remains an art form, while sex education online continues to be on the waiting list of social acceptance.

Invited for Tea

It's not every day you are invited to tea by your government. This is not the Bridgerton version of Regency niceties. In China, being 'invited to tea' is code for being interviewed, especially when you have violated specific laws and regulations.

Media scholars Jian Lin and Jeroen De Kloet argue that the Chinese internet governance rules are 'not just about "restructuring the economy", but also about restructuring culture and society'.[26] They investigated the use and regulation of Kuaishou, China's first short-video platform, founded in 2011 and nicknamed 'Snack Video'.[27] It quickly became the most downloaded app on Google Play and Apple Store in eight countries, including Brazil, Pakistan and Indonesia. Its initial popularity was due to the app targeting people from rural areas and second- and third-tier cities. Given their business model, the app relied more on e-commerce than on advertising revenue, and the algorithm rewarded content creators boasting highly engaged user bases. Chinese media praised the platform in the early years for 'revitalizing Chinese rural culture'.[28] The app gave a platform to young people from lower socioeconomic classes, who were otherwise deemed invisible.

Li Tianyou, known as Tian You, was a high-school dropout who turned into an internet celebrity by telling stories of his everyday village life. He became popular for livestreaming his rap-like performances, building a follower base of thirty-five million viewers. The honeymoon came to an end when one of his songs expressed anger at the growing wealth inequality and discrimination that rural youth like himself faced. In early 2018, China Central Television, the state-controlled network, turned against him and accused him of talking about pornography and drugs in his livestreaming performances. This quickly led to all Chinese platforms banning Tian You and ending his career overnight.[29]

LGBTQIA+ content also falls into the 'immoral' category. Creators dealing with such content are at risk of losing their platforms. The widespread internet crackdown by the Chinese state in 2021 on content deemed inappropriate, wrong or harmful pushed many platforms to act swiftly or face steep penalties. Queer groups on WeChat were erased,[30] while profiles, influencers, accounts, and entire apps were banned. Blued, a location-based dating and livestreaming service for gay and bisexual men, somehow stayed afloat. For a platform with over sixty million registered users worldwide, it is no small feat.[31] Blued positions itself as an entertainment and sexual-health company, and rarely mentions terms like 'gay' or 'LGBTQIA+', instead using 'diversity' and 'community' in reference to its user base.

However, cosmetic semantic changes aren't sufficient to stay afloat. Blued launched 'He Health' in 2019,[32] a site that allows for discreet discussions about sexually transmitted infections and other matters that can be discussed with doctors in private. The platform serves as a pharmacy where users can submit their prescriptions online. The parent company, BlueCity, succeeded in garnering a state internet hospital license for telehealth consultations. However, privacy concerns always loom when dealing with sensitive data and morality laws. A gay rights activist in China shares his concern that these internet crackdowns could inadvertently reveal users' identities, as 'some people are not really out to the public, and they don't want their sexual orientation to be revealed'.[33]

Amid these regulatory regimes, even digital humans in the metaverse may not escape unscathed. Standards are a must. The world is facing a tsunami of deepfake sexual content used to blackmail people, generate child pornography, and create misinformation and political upheaval. China is setting standards for digital assistants, virtual influencers and gaming avatars. On 28 January 2022, China's State Internet Information Office released a proposal to control deep-synthesis technology, which covers deepfakes, virtual scenes, machine-generated audio-visual content, and other emerging digital content. All generative synthetic content must be labelled, and 'destabilizing content ranging from subversive incitements to pornography to intellectual property violations' will be banned.[34]

The morality of a society will be the key driver in interpreting such regulations, from the most humane and rights-focused cultures to amplified surveillance of our intimacy and personhood.

Sexbots for Better, for Worse

Virtual assistants, digital influencers and avatars are already a multimillion-dollar industry. They are everywhere, helping people bank, shop, travel, and access sexual pleasure. Loneliness is particularly a push factor, a market opportunity for digital companions. Global searches for 'romance' and 'romantic' on Pornhub more than doubled during the pandemic, and searches for 'passionate' and 'bromance' has increased significantly.[35] This comes at a time when loneliness among the youth has risen significantly, amplified by COVID-19 lockdowns.

Decades of gender bias in the Global South have also contributed to female infanticide and a skewed gender ratio among younger age groups today. In 2018, men outnumbered women by 70 million in China and India.[36] This leaves a lot of young men lonely and vulnerable to digital scams. One such predatory app is L'amour, a Chinese app that became the most downloaded dating app in India.[37] In 2019, fourteen million users downloaded it, in comparison to Tinder's 6.6 million downloads in the same period. Indian journalists Snigdha Poonam and Samarth Bansal investigated this app and unearthed a disturbing machinery of exploitation behind it.

Figure 3.2. Fake romance economies

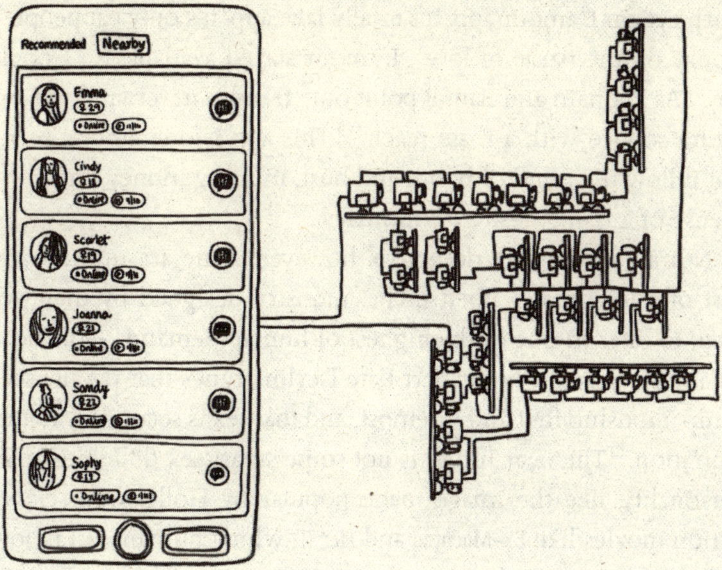

They found that ads like 'Find a girlfriend in a week' pop up on the Facebook timelines of many young Indian men. This promise of love draws many of them to download the app. Once in, they receive continual push notifications to upgrade their subscriptions so they can chat with girls and unblock supposedly incoming messages from girls. Automated messages such as 'Ayesha very interested, and she wants to talk to you', 'Mia is calling, don't miss out on her', or 'Ipshita is inviting you to a video call' inundate the young men. Some cave and pay money to chat with the pretty girls, only to learn that they have been caught up in an internet romance scam. Bots, instead of real girls, are on the receiving end of the young men's serenades.

Within months, negative reviews of this app appeared all over the internet. Despite reviews such as 'girls are fake computer control profile so don't install this app', and 'Google, please keep a sharp eye on L'amour app. It's totally fake app. It's only eat people's money on the name of love', L'amour stayed available on Google Play.[38] As Poonam and Bansal point out, 'fraud is never far from any dating service with a mass reach.'[39] This app is one among many that follow the trend of boom and bust, minting money under the license of a rising desire for intimacy.

Not all sexbots are deceptive, however. Some are upfront and part of a legitimate, albeit niche, industry designed by men, for men, to cater to one of the highest of human demands—the need for intimacy. Sex robot expert Kate Devlin argues that people seek companionship first and foremost, and that sex is secondary to that aspiration.[40] The near future is not some smart sex doll with an AI personality, like the image made popular by Hollywood science fiction movies like *Ex-Machina* and *Her* in which humans fall in love with highly advanced humanoids. Our future is of a humbler kind with the right cocktail of live-streamers, their fans and virtual stickers to express the gamut of this romance economy at a price.

East Asia has moved fast on this trend. In Korea, users took to livestreaming apps as early as 2006, and most female live-streamers engage with male fans by sharing mundane moments. These users eat, sleep, chat, dance and live their lives in front of their cameras. 17Media is one such livestreaming platform launched in 2015 in Taiwan. Within 250 days, it exceeded ten million downloads, thirty million global users, and 10,000 hours of content production daily, beating the growth rates of Instagram and Facebook.[41] The app hit a nerve as many young men move from villages to work in cities, a common phenomenon in the

Global South. A representative case is that of a middle-aged factory worker Junji Chen, who barely has a social life given his crushing work hours. This leaves him isolated and desperate to connect with his favourite live-streamer who goes by the name Yutong. While Junji has not met her in person, he expresses his affection for her through virtual stickers that can cost him a third of his salary.

These intimacy trends have undoubtedly fostered new concerns about data privacy violations by tech platforms.[42] Much of this information is deeply sensitive, given the growing number of morality laws in the Global South. Bioethics experts fear that this growing digital culture will negatively influence how women are represented, treated and related to.[43] Feminists are concerned that these digital tools will replicate toxic behaviours where conversation is porn-style and robots are treated like prostitutes.[44]

While such fears are legitimate, there must be conversations about the role of and right to sexual fantasy. Much research on sexbots remains speculative, driven by morality concerns instead of the experiences of millions deprived of basic human companionship and comfort. We have a gap between utopian visions of sexual happiness and dystopian perspectives of sexual enslavement. Media design scholars Nicola Döring, Rohangis Mohseni and Roberto Walter argue that we have a false dichotomy between humanoid artifacts such as sex dolls and robots, on the one hand, and chatbots, avatars and immersive virtual-reality pornography, on the other.[45] Dolls and robots offer the physical satisfactions of touch, stimulation, care and presence. These tactile needs are fundamental, and their value cannot be underestimated for the millions deprived of these sensorial experiences. Avatars and bots can cater to other forms of desire, tailored to the user's sexual and social experiences. By combining both affordances, we could provide the 'right degree and mixture of materiality and virtuality'.[46]

Human imagination is powerful and can fill the gap of desires to fit our needs. Sometimes, the less sensorial a digital interaction, the more intimate it can get. Companionship, compassion and care constitute much of this intimacy ecosystem. However, with increasing online misogyny, the internet may become a space more for men, by men, with sexbots as their future companions.

Digital Abstinence

The current digital culture of increasing misogyny, trolling, doxing and hate speech targeted disproportionately towards women is shifting digital absence into digital abstinence. Yet retreating from the internet is a negotiated act as the pull factors of entertainment, gossip, socializing, fashion and romance bring girls back online.

Media educator and UX research partner Kiran Bhatia spent four years in India studying how children from low-income families engaged with digital media. She, media scholar Manisha Pathak-Shelat, and I analysed her data and confronted an alarming finding—'good girls' are not supposed to go online.[47] For instance, Zenab, a fourteen-year-old Muslim girl from a conservative, religious family, was brought up to believe that digital technologies spoil girls and ruin their reputations. While she parrots the same argument, her digital behaviour reveals more nuance. She operates her Facebook account with some other girl's photo and name. She regularly comes online to 'stalk' her classmates and justifies her online presence as 'keep[ing] an eye on what others are doing'.[48]

Girls use a variety of deceptive strategies to get online. They trade their lunches for mobile top-ups, get their boyfriends to give them secret mobile phones while allowing them to track them online, push shopkeepers to share their Wi-Fi passwords, and pretend to use their limited access for homework while surreptitiously lurking on Facebook. Their digital presence takes a moralistic stance. Many girls are first in line to 'slut shame' other girls who are active on social media, post photos with 'scandalous and vulgar' clothes, or show themselves with male friends and friend them.[49] They call out girls with 'loose characters' and ban them from their social circles. It doesn't take much to get condemned. A teen girl justifies this digital policing by saying:

> In rich societies, girls can do anything, and no one will blame them. For people from my [small town in a conservative family] society, those girls are of weak character. I do not want to be like them. I want to use the phone but just to see what others are doing ... Maybe like their photos so that they know I am also on social media.[50]

Those that defy the rules of online engagement face the wrath of their peers. Another teen girl explains this condemnation:

> Why do you share your photos—obviously, you want people to see and then you blame them for seeing? Are you that stupid to share your photos online? ... Photos can be misused. Do you know what boys do with these photos of girls in their private time? Chiiii. I would never want to be that type of girl, you know![51]

Digital reticence is not unique to girls in India. My team found similar behaviours in my project on digital media use among Venezuelan refugees in Brazil for the United Nations High Commission for Refugees. Rumour mills among girls and women on romance scams, revenge porn and online harassment influence their perceptions. Rita, a twenty-year-old girl in the settlement, reflects on why she thinks twice before posting anything on her timeline: 'That friend of mine who gets a lot of likes uploads photos in a top, in lingerie. But I'm not going to do that crazy stuff, because there's a lot of harassment that comes with that. There is a lot of harassment on social media.'[52]

There is pride among many girls in being offline, as they see themselves as upholders of their families' virtue and their communities' honour. Yet they yearn to be seen by their peers and want to feel part of this digital social network. These girls view it as a smart move to be offline, but find it emotionally hard to resist being online. They live their online lives in a space between digital demonization and desire.

Safe Space for Romance

Tech entrepreneurs in the Global South recognize the glacial pace of cultural change and some provide alternative ways to participate online. Soulgate, an 'algorithm driven virtual social playground where people can create, share, explore, and connect' is one such experiment.[53] With more than thirty million users in China, this app connects young people with similar interests using algorithms, but with one major difference—they can only communicate through avatars. To be clear, Soulgate is not a dating app: it does not aim to build offline relationships for its user base, and so it does not allow its users to use their real names, photos or voices. The platform's purpose is for young people to drop their labels of job, income, status and gender and connect with others like them, to 'have no lonely person in this world'.[54]

Anonymity is meant to deliver the freedom to connect and experience digital intimacy. In the last decade, despite good intentions, many apps using anonymity produced increasing sexual harassment online and privacy violations, resulting in their bans. A classic case was the ban of the Secret app in 2014, due to the proliferation of nonconsensual sexual content using their anonymity feature.[55] A more recent case is that of Telegram, an end-to-end encrypted chat platform, that came under scrutiny in 2022 for allowing the sharing of women's intimate pictures without their consent. The BBC conducted an investigative report on this matter that revealed how this platform's minimal content moderation and anonymity resulted in the nonconsensual sharing of thousands of photos and videos of women in at least twenty countries.[56]

Some tech entrepreneurs see the bias towards visual content in innovation as the core obstacle to intimacy. FRND is a dating app launched in India in 2019, and targeted at second- and third- tier towns. With over five million registered users in 2021, FRND claims to provide 'a safe, secure and fun environment for boys and girls to chat anonymously via audio streaming' with a goal of creating a 'dating ecosystem for the next billion internet users'. Further, FRND does not allow real pictures, only AI-generated avatars. Their business model utilizes virtual gifting to monetize online courtship between strangers. Bhanu Pratap Singh Tanwar, CEO and co-founder of FRND, explains that by providing a safe platform through public anonymity for Indian youths, especially girls, companies can profit from this untapped romantic market.[57]

Privacy-driven apps have become synonymous with dating apps. However, building designs based on fear and shame alone would be a disservice to one of the most important joys for human beings—sex. Some organizations are taking the lead to shift their approach from privacy-driven to pleasure-based engagements. Anne Philpott, director of the Pleasure Project, an international education and advocacy organization, puts 'the sexy into safer sex' by promoting pleasure-based education. Her team has analysed studies on such interventions over two decades and found that being pleasure-centric not only adds fun to our social equation but also has positive impacts on condom use, safe sex, mental health and happiness.

What we need are designs for desire that centre care, compassion, touch and intimacy, especially at a time when our lives are becoming increasingly remote and lonely.

4

Surveillance of Care

'LOOK AT THEM! I THINK THAT'S REALLY WHAT THE ESSENCE OF dancing is about—not showing off but really bringing happiness and joy to the dancers themselves in a simple way,'[1] commented someone on a viral video of a couple dancing in rural China. The couple in question are corn farmers Peng and Fan from the countryside of Ruian, Wenzhou, who took to dancing when faced with personal adversities of financial stress, health issues and depression. They invented the 'rural shuffle' dance, which was watched and praised by millions in China and abroad. Their positivity struck an emotional chord among people, tapping into a contemporary yearning for purity, simplicity, resilience and joy.

More than a hundred years ago, French sociologist Émile Durkheim coined the term 'collective effervescence'[2] to describe shared human sentiment. He postulated that collective emotions built through everyday acts of sharing stories, gestures, movements, and other humble forms of expression can lead to social solidarity, bonding, and a sense of belonging. Today, memes, hashtag cultures,

WhatsApp groups, YouTube channels and Facebook pages have become cultural and cathartic repositories that can bring people together on matters from the mundane to the profound. Zoom, in a way, got us through the COVID-19 pandemic. While we seek each other online for solace, strength and support, social surveillance using digital tools comes with a dark side. Collective effervescence is fodder for the capitalist machine as our social lives become data to be traced, tracked, and turned into value for states and brands. As social psychologist Shoshana Zuboff argues, we have become 'means to others' ends'.[3]

Is there a way to steer social surveillance towards tools to build social health and collective well-being? To rebuild social trust, society needs a surveillance system of care—one that moves away from watching each other as a form of policing to watching over one another as a form of recognition and compassion. The culture of fear needs replacing with a culture of collective empathy. Memes can boost mental health. Open-source data and crowdsourcing can build maps of neglected areas and generate empathy through visibility. Refugees and their families can track each other for safety and security. Female workers can organize themselves digitally and protect themselves by rating unruly customers. Sentiment analysis of global emotions can help improve public policy to increase global happiness. Digital tools can serve us by feeding the collective effervescence we desperately need in our societies today.

Eyes on the Street

To build a safe public space, there must be eyes on the street.[4] In the 1960s, US urban reformer Jane Jacobs introduced this idea as a useful rubric when approaching city planning. She argued that neighbourhoods would become safer if buildings faced the street. People who live, work and inhabit these public spaces can look out for the safety of both residents and strangers, although it is a thin line between watching each other and watching *over* each other. To deter care from becoming a means of control, power brokers need to be distributed, decentralized, yet connected to those that can institute change. This network formation demands a democratic crafting of the social order. Jacobs remarked, 'The trust of a city street is formed over time from many, many little public sidewalk contacts ... Most of it is ostensibly trivial but the sum is not trivial at all.'[5] This principle applies to our digital spaces as well.

'Our main demand is not to get killed,' explains Camila, a student in Mexico City, on why she joined a Facebook and WhatsApp women's group to track each other's real-time locations.[6] Femicides, the gender-biased killing of women by men, rose by 135 per cent in Mexico in 2021 and constituted a quarter of the 3,750 women killed. While more than 90 per cent of people in Mexico have Facebook, the majority find the interface boring. Yet for the purposes of safety and solidarity in the face of growing gender-based violence, women appreciate the platform's simplicity and functionality. They share their stories, alert others to their plight and reach out to one another for help through this channel.

During the COVID-19 lockdown, mothers in China looked out for one another and boosted each other through humour. The #workfromhomewithchildcare meme in China received millions of posts, likes and shares.[7] Mothers used emojis, animations and word art to mock expectations to balance work and childcare when schools shut down. Profiles on Chinese TikTok like that of Chubby Toot, the username of a twenty-five-year-old Chinese mother, attracted two million followers as she shared her struggles when working at home with her two-year-old son, whom she refers to as her 'little rascal'.

Solidarity surges have spread worldwide. Media researcher Maud Ceuterick calls humour 'a great feminist weapon'.[8] *Atlantic* points out that 'humor is everywhere because fear is too'.[9] Youth solidarity has also increased worldwide, especially in the Global South, where young people face crushing workloads, social pressures and family responsibilities.[10] In recent years, the 'lying flat movement' of young Asians refusing to work long hours has grown. Many youths view their existing workloads as untenable and even undesirable. In July 2020, rural Chinese vlogger Li Ziqi set the record for the most subscribers of a Chinese language channel on YouTube, with more than sixteen million subscribers.[11] Videos of her pickling vegetables in the countryside, making a dress out of dried grapes, horse riding in the forest, and building a thatched roof have inspired youths from Bangladesh to Portugal.

We tend to underestimate the value of ordinary acts from strangers whether they be sharing a joke, saying a kind word, or being decent with one another. Canadian journalist Malcolm Gladwell argues that 'kindness [from strangers] is a temporary suspension of indifference'.[12] These acts can serve as a social glue, holding humanity together.

Digital Detox

Pushing against anti-social behaviours takes collective effort, with help from influencers, users, civic groups, regulators and algorithmic innovations. The Goodness Bot (@goodnessbot), an automated X account, when tagged in a reply to a bullying tweet, generates a similar version of the offending tweet, but replaces the negative messages with kind and humorous words.

US activist Monica Lewinsky, in partnership with the advertising firm BBDO, spearheaded this initiative as part of a larger anti-cyberbullying campaign. Lewinsky taps into her past of being publicly bullied and how strangers' kind words got her through:

> There was a time where getting the mail was the highlight of my day ... so I know the power and the impact of hearing from a stranger ... That's what this bot can do. You don't have to know the person. It's just passing along the good deed in a way and trying to help someone that you see who may be in pain.[13]

TikTok's algorithms have come under scrutiny as they are accused of privileging content such as body shaming that is harmful to minors. The scale of outreach is staggering. In 2021, posts with a body-shaming hashtag on TikTok got 1.2 billion views.[14] TikTok tries to reduce these harms by creating blocked lists of keywords and profiles, which has limited effect as content creators and users circumvent these blocks by deliberately misspelling keywords to evade filters.

To tackle this crisis, body-positive influencers have come together as a counterforce. Organizations like Within Health, a digital health startup, suggest a social and technical strategy to fight eating disorders. Crowdsourcing keywords and associated songs, images and themes from users helps keep the blocked list dynamic and responsive to ongoing changes in how people communicate body shaming when online. TikTok, however, can prime mindsets by disabling harmful content through their autocomplete, ranking and search functions. The company can provide users with opt-out choices on what they consume. They can deploy trigger-warning labels, embed pop-up tutorials to bring awareness to a problem when users search for negative content and collaborate with other platforms to standardize this impact.

Another initiative comes from Line, one of the most popular encrypted messaging apps in Asia. Line launched a programme to tackle misinformation in Taiwan in partnership with local fact-checking organizations like Taiwan Fact Check Center and Cofacts.[15] The app allows users to forward messages they want verified and receive an answer in real time about whether the content is true. This seemingly humble intervention allows the company and other stakeholders to tackle harmful content without breaking end-to-end encryption. The novelty of these digital tactics is less in their technological innovation than in social relationships and partnerships to build 'natural proprietors'. Such tactics, while trivial by themselves, can cumulatively nudge users toward a cultural shift towards collective care online.

Off Switch

Exercising collective care does come with challenges, especially when caught in the age-old struggle between freedom and control. X indiscriminately purges bots to build trust. The quality of fact-checking is contingent on what kinds of training human and AI moderators get. Blocking keywords can inadvertently censor legitimate content, including political critique of the state, in the name of social harmony. The freedom-control debate gained renewed vigour during the COVID-19 lockdowns as states used AI-enabled cameras, digital apps, heat sensors, mobile tracking and social media content to enforce public health measures.

Internet governance scholar Laura DeNardis argues that the internet is no longer just a digital public sphere within which we connect and communicate. It is an infrastructural system that has 'no off switch'.[16] This system is both virtual and physical, connecting vehicles, wearable devices, home appliances, drones, medical equipment, currency and every other tool that mediates our social life. Ubiquitous social and data surveillance, however, is far from a reality for the majority of the world. While there may not be an 'off switch', there are loose connections, poor bandwidth, VPNs, offline sharing of profiles and devices, hacking, weak digital literacies and other factors. Being 'on' 24/7 is a myth for many.

The surveillance of the world is fragmented, unrepresentative and faulty. While the omnipresence of surveillance tools sounds ominous, those at the margins have long been invisible, silenced, unmapped and undocumented. In some cases, digital surveillance provides security that enables freedom. Senior humanitarian advisor Ivan Gayton of Humanitarian OpenStreetMap says, 'There are about two billion people in the world who don't appear on a proper map.'[17] A third of humanity will live in slums or informal settlements in the future.[18] The growth of slums accounts for 90 per cent of urbanization in the twenty-first century, yet these slums are largely unmapped. Smart cities of the future will resemble Kinshasa and Mumbai, not London and New York.

Having a digital identity matters. More than a billion people with no formal identity want to opt in, not out, of this state tracking system. Digital identification would help them open bank accounts, secure funding, apply for jobs, exercise their legal rights, enable formal claims, and prevent undue detention and deportation.[19] For instance, while India has made strides in economic growth in the last few decades, 90 per cent of workers remain informally employed as of 2022.[20] Many of the vulnerable workers are young, female, and belong to groups of migrants, scheduled castes, scheduled tribes and other 'backward' classes. As surveillance scholars Murakami Wood and Rodrigo Firmino argue, 'The fear of anonymity and being "lost" is far stronger than any concern about surveillance or control.'[21]

Tracking people, supplies, infrastructural projects and access to welfare schemes can lead to more accountability and fairness in the distribution of scarce resources. This applies particularly to the shadow economy in the Global South, where precarity is the norm and informality is the way of life.

Feminist Futures of Work

Figure 4.1. Collectives and feminist design

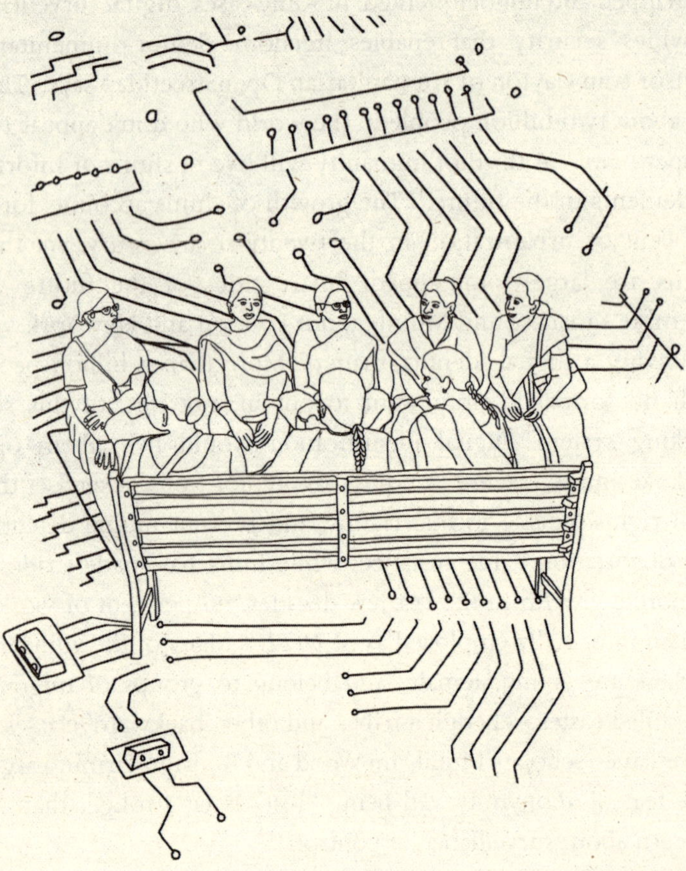

At the onset of the pandemic in January 2020, I founded FemLab, a feminist futures of work initiative, with feminist media scholar Usha Raman. We received funding from the International Development Research Centre (IDRC), a Canadian grant agency. This project was part of its larger funding initiative on digital innovations and the future of work in the Global South.[22] IDRC focuses on how automation and digitization impact the economy in the Global South, with approximately sixty million workers engaged in sectors transformed by digital interventions.

FemLab focused on the changing work conditions for female gig workers in India and Bangladesh during the pandemic in the ride-hailing, home-based salon services, sanitation, artisanal and construction sectors. We built a team across four cities in India to examine how these workers were organizing themselves using digital tools in ways that could boost their statuses, voices, opportunities, conditions and livelihoods. We wanted to unpack the surveillance of care apparatus in the world of work at the bottom of the value chain.

The project came at a time when remote work became the norm in many sectors during COVID-19 lockdowns. Workplace surveillance technologies boomed around the world, including in India. While venture capital funding was down for most sectors, remote monitoring tech received more than $394 million, a 62 per cent increase from 2021.[23] Companies wanted to comply with health measures while ensuring workers' productivity. Informatics scholar Jessica Vitak has explained a cultural shift among employers who believed that having a 'treasure trove' of data on their workers could lead to making them more productive.[24]

The Centre for Internet and Society (CIS)[25] reported the exponential growth of monitoring technologies such as facial-recognition softwares, human-efficiency trackers, keystroke loggers, and social media screeners being used on Indian employees.[26] CIS found these software running in the background, collecting data from employees' WhatsApp, email and social media accounts, even past their work hours.

While there are data protection policies in place to curb companies' indiscriminate collection of their customers' data, there are few comparable measures to protect employees. The choice can be stark for workers. Employees consent to this intrusive surveillance because they fear losing their job. This power asymmetry is far worse in the informal sector, which constitutes the bulk of India's economy and is increasingly becoming digitized.

In 2020, sanitation workers in Chandigarh, in India, protested having to wear GPS-enabled watches that tracked their movements. The municipal corporation (MC) officer justified the use of these wearable devices by saying, 'We often get complaints that MC sweepers, gardeners, sanitation inspectors and enforcement staff are absent from field duty. This new system will rein this in.'[27] However, many workers perceived this as 'bonded labour', given that the majority of the five million sanitation workers come from the marginalized Dalit caste.[28]

The reliability of these tools also came into question. Women sweepers wearing bangles complained that their watch would turn on when their bangle accidentally touched the watch. Some complained that the watches showed the wrong location, causing a loss of wages. Their salaries could decrease if the wearable's batteries died while they were at work, failing to capture their time.

Few redressal mechanisms and high demands on the workers' time to address glitches result in more precarity and exploitation. As the private sector experiments with new tracking and automation of their workers, the public sector experiments with new ways to deliver essential safety nets, albeit with mixed results.

Online Safety Nets

In 2021, the Indian government launched the e-Shram portal, a national database of unorganized workers. The portal registers unorganized workers and connects them with social security benefits if they meet with an accident, become disabled or unemployed or die at the work site. This is an unprecedented effort to formalize the massive informal economy.

Sociologist Shweta Mahendra Chandrashekhar worked for FemLab with a focus on the construction business in India. She interviewed union workers and leaders, municipal officers and contractors to understand worker benefits, including how digital registrations improved access to welfare benefits. Her father owns a tunnel construction company in Pune, and her childhood was shaped by her visits to the construction sites and her interactions with the site engineers, migrant workers, contractors and government officials. The COVID-19 lockdown hit the construction sector particularly hard since a majority of their workers are migrants. Pune saw about 75 per cent of the workers leave for their villages when the lockdown was announced overnight in March 2020.[29]

A majority of migrants failed to access welfare benefits, despite the government making funds available during this humanitarian crisis. Union leaders provided reasons for these failures: migrants couldn't directly register online, so contractors needed to register workers on their behalf. Sometimes the contractors themselves were unregistered and were subcontracted by other private contractors. Middlemen have few incentives to register migrants, as laws such as the Interstate Migrant Workers Act demand that they provide migrant workers and their families with healthcare, housing, education for children, semiannual train fare to their villages, and other benefits. The president of the Construction Workers Union, while hopeful about digitizing the registration system, argued that we need to simultaneously build awareness of these schemes and provide free digital literacy support to enable online registrations. He remarked that otherwise, the digital tracking process could create another layer of middlemen and more opportunities for exploitation.[30]

The government's efforts to standardize registration and simplify migrants' access to benefits across states are fraught with challenges. These goals require consolidating migrants' registered accounts, including their pension and other accrued benefits. The idea of a singular professional identity doesn't match the reality of migrants who are seasonal farmers and work simultaneously in construction, ride-hailing and domestic services to make ends meet.

Women face similar issues of being absorbed into the state welfare system. Their situation is compounded by the patriarchal culture in which being too public, too outspoken and too present can have adverse consequences. Monitoring apps can amplify the social pressures women face and further deter their participation in the workforce. According to the World Bank, the percentage of women working in India decreased from 2010 to 2020, dropping from 26 to 19 per cent.[31] In fact, this decline has become a global trend, especially among female gig workers. Women are increasingly abandoning the use of digital tools and the internet itself as fear of a digital presence overtakes them.[32] Locked profiles, while an option, keep them from capitalizing on the attention economy. Tracking devices deter movements. Social expectations relegate them to care work, which remains undervalued and underpaid. In such an ecosystem of withdrawal, what will it take for women at the bottom of the supply chain to go online and use networked tools to their advantage?

Not Just a Woman

Figure 4.2. Women behind the wheel

FemLab' s media researcher, Pallavi Bansal, and I designed a project to gain insights on why it was so difficult to find female drivers for ride-hailing companies despite the high demand. We found double standards for women working with strangers in public spaces. Sushil Shroff, director of Taxshe, an exclusive all-women driver-on-demand cab service running in Bangalore, explains that cultural norms result in female drivers being judged harshly and can deter them from choosing this as a profession. Shroff explains that there is a tendency to judge women drivers as 'being cheap and available', because 'anyone can sit in her car and talk anything to her, and she'll have to take it'.[33]

Taxshe had to rebrand women drivers as 'alternate moms' to attract women to this profession, get their families to agree to them working in the ride-hailing sector and shift customers' mindsets.

The company leveraged the cliché trope of caring women that feminists typically want to break out of. They labelled female drivers as 'Roos'—referring to kangaroos—who will protect young ones while moving. Shroff explains that they had to try to convince clients to respect drivers by equating them to mothers: 'We told each of our clients, "like you, she's a mother. Yeah, she's handling her kids, and she's driving your kids ... It's a parent who's driving. It's not just a woman." ... So that gets respect.'[34]

Non-profits like the Azad Foundation[35] and the Neeva Foundation[36] explain that it is essential to invest in female drivers to prevent them from dropping out. Women may have never driven a car or even opened a car door for a customer. They may not be used to talking with strangers. Filling out forms, getting a license, dressing professionally, handling finances, and dealing with the mobile app takes time and, more importantly, confidence. One driver, a widow with two daughters to support, shares her story:

> I was very afraid of starting new things ... my confidence level was not that good. First, they give you the confidence over there. Like 'You can do, you can do', that one is the first thing ... I was so afraid of everything ... I hesitated to speak, I hesitated to approach people, like that. That is one thing I am not now.[37]

While these organizations build women's confidence and capacity as drivers, some fixes are more related to urban planning, social norms, digital affordances, awareness and media literacy.

Empowering 'Partners'

Women drivers at Ola and Uber express frustration about the lack of access to clean and safe toilets. Ola Mobility Institute Foundation, a think-tank focusing on mobility innovation in India, explains that it is also the lack of awareness of app options that contributes to this plight. It conducted surveys among women drivers and discovered that few were aware of how to navigate their app to find the nearest restroom.[38] Additionally, the app can benefit from a rating system on the cleanliness and safety of restrooms for women. The SOS feature on the app, while present, is rarely connected to a live operator and often results in no response. Women drivers want to be able to rate their customers, too, and have their company aggregate these ratings. They want the company to act on the data insights and block customers who broke rules of engagement with persistent rudeness, harassment and overall bad behaviour.

Women workers or 'partners' with home-services gig platforms like Urban Company shared similar concerns and desires for the ratings to be a two-way process with our FemLab researcher Sai Amulya Komarraju.[39] Moreover, while they recognize the risks of everyday digital surveillance, they also see in them an opportunity to prove their worth and distinguish themselves. They feel pride in completing the ten-day training course, given that some fail in the onboarding process. They learn how to sanitize their products, deliver their services, present themselves, greet customers, operate the app, and check their commissions and ratings. The women learn the vocabulary of this digitized sector—power bank, app, leads, credits, ratings, reviews, automated, tracking, app store. These terms become part of their standard operating procedures. The partners feel professional and respected, and the data proves their standing to the company and customers alike. A service provider confidently explains the rating system:

> Now you [the customer] will be questioned, how was the pro who was sent to you, did she behave well, did she sanitize, wear PPE kit, gloves, mask. If you see me doing all of this, then you will tick yes, but if I have never used any of

> these things, you have never seen me use these things you say no.[40]

Many feel pride in being able to manage expectations, comply with hygiene standards, and maintain a high average rating in the system. Monitoring and rating systems can be repurposed as tools for transparency, company negotiations, promotions and social recognition. The professionalization can have a spillover effect on the standing of these women workers in their community:

> The moment you say beautician, to be honest, the mindset is they look at you differently, not with good perspective or good intention. Of course, this does not mean that everybody is like that. But UC [Urban Company] is a brand, so in the starting, during the training itself all of this is clarified. If you are working in UC, then people should automatically differentiate between you and whatever preconceived notions they have about beauticians; they must perceive us as thorough professionals. There is clarity about this now for sure.[41]

Building surveillance of care requires negotiating deep-seated cultural systems and infusing power into given labels such as 'partners' by appropriating tools at one's disposal. However, this is done through soft power, subtly increasing agency from within, and starting with a core power unit—the joint family.

Family Planning of a Working Kind

Meesho, the reselling app, learnt this cultural norm quickly as it scaled its business model. Engaging women resellers on their platform is a family effort. Achyutha Sharma, the company's user research manager, explains how over time these compliances lead to subtle but important shifts in the women's status, freedom, and confidence in and outside the home. Sharma points out that at the start a woman must negotiate 'social permissions' with her husband or family decision-makers to get a mobile device and internet access for work. The family typically agrees on the condition that she also fulfil her household duties. Over time, as they see profits coming in, things begin to change in the woman's status in the house and her bargaining power, leading to a higher degree of 'digital confidence'.[42]

Family surveillance can ease over time to a healthy inattention that allows women to become more visible and vocal online. These tactics, while commendable, underline the weight of the patriarchal apparatus women need to push against to have a fighting chance in an increasingly automated workforce. The future of work will continue to be upended by disruptive technologies. Workers at the bottom of the value chain bear the heaviest burden. This is especially true of women in the informal sector. Media debates position labour futures as a contest between tech and humanity.[43] Tech, however, *is* human. The digital is physical, built on the continuing sweat of workers' experience. Algorithms learn from people as much as people learn from them.

Building institutions, infrastructures and initiatives of care such as clean and safe toilets for women, two-way rating systems, confidence training, upskilling, and universal basic incomes can sand down the hard edges of patriarchal norms. These changes start from a position of compassion. Installing the surveillance of care demands a shift from the rational to the emotional. Feeling can become fuel to action, unless it is commodified against those that need the most change. Machinery and empathy do not have to stand against each other in this pursuit of social change.

Emotion 3.0

Machine learning is built to identify and structure opinions, emotions and beliefs from collected text, expressions and other biometric cues through wearable devices. Businesses valorize emotional intelligence (EI) to help build relationships, manage crises, communicate effectively and inspire their teams. It provides a competitive edge. A 2019 survey with managers in the United States showed that 71 per cent of employers favour the so-called emotional quotient (EQ) over the intelligence quotient (IQ) in employees.[44]

Empathy and social skills trump logic and problem-solving as management gurus tout EI as the key to good leadership. Impression management is all the rage. Recruiters are increasingly giving weight to EI when hiring.[45] Insurance plans, advertising campaigns, product designs, banking protocols, healthcare tools, education content, and agricultural innovations are increasingly tailored to their diverse customers by mining their sentiments. This outlook was far from the norm in the past: emotionality was perceived poorly in society and was associated with women and the Global South. To be emotional was to be weak.

Personal Polygraph Machine

Feminists have long battled the undermining of emotion and its link with the feminine to keep women away from political life. A persistent trope of women revealing their emotions easily, like a 'personal polygraph machine',[46] has worked against them, as leaders are expected to be self-contained and self-controlled. This degradation of emotion applies to nations as well. The international peace organization Service Civil International, in their 'Picturing the Global South' toolkit, highlights how countries in the Global South are typically perceived as 'emotional' by those in the North.[47]

In the colonial days, imperial settlers framed Southern cultures as 'emotionally pathological'.[48] They viewed the habits of their new subjects as the antithesis of the Enlightenment values of reason, scientific progress and technological order—the makings of masculine sensibility. Their worldview provided a 'rational' justification and moral responsibility for settlers to 'civilize' the natives. Anthropologist Ann Laura Stoler explains how colonialism was driven by an 'emotional economy', where rulers strove to predict and prescribe what sentiments to perpetuate and what emotional contagions to contain.[49]

Until the late nineteenth century, emotion connoted more a collective than an individual state of affairs. It alluded to groups driven by feeling to action; as gender studies scholar Jacqueline Holler explains, people were 'powerless to resist'.[50] Inspired by the church and politically vested interests, emotional standards set in to manage and control people's sentiments through rituals, institutions and cultural practices. Settlers validated certain types of emotion, steered by beliefs about patriotism, racial superiority, paternalism and honour.

Today, as emotional intelligence meets artificial intelligence, how do we negotiate the potential of collective care and the fear of collective control? As Stoler argues, 'Sentiment is the ground against which the figure of reason is measured and drawn.'[51]

Age of Sentiment

Cultural anthropologist William M. Reddy recognises emotion as the driving force of the Renaissance, the 'age of reason', and retitles the era the 'age of sentiment'.[52] He may have inadvertently captured our contemporary time. The battle for the future of our sentiments as an instrument of care versus control has begun.

Meta has been inundated by controversies over its experiments using sentiment analysis. Studies conducted as early as 2014 by their Core Data Science Team and Cornell University's Departments of Communication and Information Science found that 'emotional states can be transferred to others via emotional contagion, leading people to experience the same emotions without their awareness'.[53] In June 2022, Microsoft announced its new policy to phase out public access to several of their 'emotional recognition' tools (such as Azure) after criticism of their flawed results.[54] Experts argue that without accounting for context, facial expressions cannot be equated to internal feelings.

Spotify appears to have 'solved' some of the context issues with its 2021 patent on monitoring users' speech to infer their tastes.[55] The AI captures intonation, stress, rhythms and other elements of speech to assess emotional states. Machine learning contextually situates these insights with identity-based metadata (such as gender, age and other such variables), and environmental metadata of users' locations and physical surroundings to get a more accurate analysis of the emotions driving music choices.

During the COVID-19 pandemic, Zoom became a living room for billions of people, and a global social experiment on people's emotional analysis. Zoom analyses users' facial expressions, vocal patterns, body language and gestures to deduce collective human behaviour. In May 2022, Fight for the Future and twenty-seven other human rights organizations wrote an open letter to Zoom, calling for the halt of their AI-driven emotion tracking software.[56]

While the efficacy of sentiment analysis continues to evolve, the intent to weaponize collective emotions remains a serious concern. How can regulators break away from these binary choices to steer tech toward a care-based approach? Can tech companies reverse-engineer these tools to benefit society?

Reverse Engineering for Care

Tech companies must be pushed into action. Meta can build on such analytics to reduce sexual harassment, foster mental health and build community. Spotify can repurpose itself as a music therapy portal and provide comfort to millions with better targeted playlists. Microsoft's Seeing AI, a tool that helps blind people record their experiences and events, can be enhanced with sentiment analysis to deepen users' recorded memories. Zoom can continue to build on its living-room feeling by designing for contagious joy, shared grief and solidarity, as we witness online weddings, funerals, birthdays, graduations and other events that mark our lives.

Take the field of medicine, supposedly the most rational of fields. Clinical practitioners are trained to view emotion as obstructive to care. 'Hysterical' patients need containment. There is little space for depression, anger or faith in decision-making. In recent years, medical experts have started to question 'objective' care as they find evidence that some providers offer discriminative care treatments based on race, class, gender, age and ethnicity. Health sociologist Amanda Gengler pushes us to elevate emotions in healthcare to make better decisions and choices of care.[57] Emotions, when imbued with symbolic value and mobilized in the decision-making process, could make decisions easier to arrive at, increase feelings of confidence in decisions made, and stave off interpersonal conflict.[58]

Understandably, however, attending to emotion is difficult and exhausting. Nurses and doctors continue to suffer burnout. Our current global healthcare system is under tremendous strain, which is not sustainable. Sentiment analysis can play a part in reducing the burden of care among providers. These tools can assess patients' states of mind and generate messages that can best induce the most responsible behaviour.[59] AI analytics of social media content can capture anger, angst and fear related to topics like vaccines and help craft more nuanced strategies for outreach and awareness. Digital surveillance systems and prediction models can conduct real-time analysis of public emotion with disease progression in relation to demographic and geographic contexts. Sentiment-mining techniques could become one of the most critical tools for public-health officials in designing evidence-informed policy.

While health systems are having their 'emotional' moment, and tech companies are pushed to recalibrate for compassion, self-care apps are reaching out directly to users. Personalized care is built on mass emotion and speaks especially to users who have long felt forgotten.

Self-Care in the Wild

Apps in one sense are bandages on a global need for accessible and formal mental health programmes. Simultaneously, apps are more immediate market responses towards people's needs, which could stimulate the conversation and policy on better institutional mental health access for all.

In the last decade, mobile health (mHealth) apps have skyrocketed in user uptake, partly due to billions of Global South users coming online. The mHealth market is booming globally, valued at $38.2 billion in 2021 and expected to expand at a compound annual growth rate of 11.8 per cent from 2022 to 2030.[60] These apps span the full spectrum of self-training, symptom management, information seeking, risk assessment and home monitoring. While the tools drop the human intermediary, sentiment analytics-enabled chatbots can introduce a humanizing dimension and improve the digital bedside manner.

Global South millennials are leaders in app-ifying their emotions. China, Turkey and Argentina make up three of the top five countries generating self-care app revenue in 2022. If we are to go by today's self-care market broadly, the emotional economy is not just lucrative, but also empowering. Emotion is now at the centre of a growing genre of self-care apps that are especially popular among millennials. Moodpath helps millennials track their moods for daily self-reflection. I AM is a daily affirmation app to perk up youth during their day. Virtual Hope Box provides games, coping cards, feel-good visuals and soothing music to help users deal with stress. Calm helps you meditate. Meditation apps alone were projected to earn $3.71 billion in 2022.[61]

Bots have become increasingly anthropomorphic as emotion enters algorithmic solutions. AI companies have added feminine features to build comfort and trust among their users.[62] People are increasingly assigning qualities like warmth, friendliness, empathy and communal feelings to AI tools. Consumer ethicist Sylvie Borau and her team point to an ethical quandary that AI designers and policymakers face. They find that gendering tech helps improve consumer uptake of apps. However, they argue that if we destigmatize emotion and elevate it as a critical asset, we could transform the gendering of design to be empowering, positing that feminizing AI could 'inject a unique human essence into such lifeless, inanimate tools'.[63] In turn, it could lead to some form of customized care where AI lends a helping hand.

Coded with Care

Mental health is a pervasive concern as our lives are increasingly digitized, automated and alienated from others. Social well-being is of paramount concern as the fabric of emotion is tearing from divisiveness of all stripes. Emotion runs high.

Gallup, a US data analytics firm, conducts an annual poll on global emotions to reveal the emotional state of people in more than 100 countries. The 2022 report[64] showed that people have become significantly unhappier over the years and that one-fifth of all adults do not have a single person they can count on for help. War, poverty, loss of work, environmental devastation, political instability and loneliness have made us more miserable. Riots, strikes and demonstrations increased 244 per cent from 2011 to 2019. Negative emotions make people crueller. Depression can lead to death. Loneliness can make us bitter. The report concluded with a push to rethink our detachment from one another, as a 'world filled with negative emotions make people behave differently'.[65] Negative emotions beget negative decisions and destructive actions, which spread faster than a virus, especially when amplified by social media.

Moreover, while people, especially in the Global South, struggle to get jobs, fulfil their aspirations, or find more humane work environments, politicians in the West are increasingly expressing anti-growth sentiments. *The Economist* marked the end of 2022 with a reflective piece on how the West has fallen out of love with growth and has distanced itself from free-market ideals.[66] This polarizing worldview is further widening the North–South divide. The anti-growth sentiment driven by an aging Western population can stifle the collective aspirations of young populations in the Global South for a better life, more innovation and serious reforms. What if tech companies could channel this angst into productive care by ensuring our digital tools work for humane growth?

Emotion needs rebranding. Let us infuse it with the meaning of joint action. Collective emotion should be neither romanticized nor denigrated. It is a tool much like any other that, and when used to our advantage, can inspire decency, kindness and fairness. Like all tools, it requires a village of stakeholders to become a success. Algorithms, when closely audited, monitored and accounted for, can be a friend to whom we unburden ourselves of the weight of social expectations, unbearable routines and crippling loneliness. Claims made by apps, tools and devices, however, need humbling.

An emotive nudge can produce a ripple effect to meet a need. The trivial adds up and can create a cultural shift. The marginalized recognize virtue in marginal change, coming from the inside, slowly, stealthily, but with cumulative impact. Women working within conservative family settings and informal workers in precarious jobs can draw strength from the collective anonymity of emotive analytics to signal change.

The surveillance machine, if designed as envisioned by Jane Jacobs, can help us take care of one another. Users can do their bit by ensuring they don't stay silent before online toxicity. Designers must figure out how to repurpose virality, attention, personalization and other data tools at their disposal for nurturing our mental, physical and social health. Politicians need not throw the baby out with the bathwater. They can use policies to build competitive standards of care and compassion. To code with care is to design for context. The growing global climate crisis, loss of biodiversity and environmental degradation demand an extension of our social care apparatus to our planetary well-being.

5

Green Design

THE DESIGN COMMUNITY IS INCREASINGLY NORMALIZING THE idea that what is good for the planet is good for the people. The American Institute of Architects has made a 'materials pledge' to align their designs with values that support the planet and promote social equality. They are developing metrics to translate these values into standards[1], seeds for a paradigm shift that were sown a decade ago.

In 2012, the European Commission reported that 80 per cent of the environmental impact of products is determined at the design stage.[2] The United Kingdom Green Building Council sounded the alarm on the construction industry being responsible for 25 per cent of total UK greenhouse gas emissions,[3] and the Ellen MacArthur Foundation reported that businesses could reduce global greenhouse gases by rethinking how they design everyday products.[4] In the last few years, podcasts, pledges and proposals have sprung up among diverse design communities to get on board with this paradigm shift. This time, the money has followed. The 2022 Climate Tech report from Tech Nation revealed that climate tech investment raised an unprecedented $111 billion by start-ups and scale-ups globally.[5]

As early as 2016, the International Finance Corporation promised $23 trillion to the global agreement on climate change adopted in Paris, seeding climate-smart investments till 2030.[6] This came at a time when the world's majority outside the West was paying a disproportionate price for decades of Western consumption, and still is. United Nations Secretary-General António Guterres reminds us that 'As is always the case, the poor and vulnerable are the first to suffer and the worst hit'.[7] Environmental activists in the Global South call for climate justice to equitably redistribute the burdens of climate change. Yet, European policymakers and activists rarely look beyond Europe's fortresses to pioneer green regulations that will impact the world. They see themselves as the world's regulators by establishing benchmarks, guidelines and standards for the rest to follow. The 'global' value chain starts and ends at their borders. US designers may embrace the rhetoric, but not the practice of sustainable living of many at the margins.

'America first' often translates to 'America only'. Corporate speech continues to define the billions of people living outside the United States as 'rest of world'. In these circles of power, the 'global' may be an inconvenience, a threat, high risk, or a weight that they must carry. To do things differently, we can learn from planetary care practices among indigenous communities in the Global South. Striding forward doesn't always mean leaving the past behind. There is value in culture for computing a future. We need to dive deep into diverse cultural contexts to avoid reinventing the wheel. Sustainability is indigenity writ large.

Design for Planet

In 2021, design formally went green. The Design for Planet Festival was the first of its kind to bring together designers, architects, businesses, governments and activists to rethink how to design our way out of the climate crisis. The festival was part of the prominent global climate summit known as the COP or Conference of the Parties. The United Nations has been hosting this global gathering of world leaders for three decades, but what started as fringe interest for the planet's well-being, has now reached centre stage in the global arena.

I was invited to give a keynote at the festival, in the second week of November 2021. This was going to be my first trip outside of the Netherlands since the COVID-19 pandemic struck. Between the antigen tests and the tram, train and overnight ferry into Brexitland, travel had lost the allure it once had. However, as soon as economist Kate Raworth kick-started the festival by asking, 'If you're not designing for the planet, what planet are you on?'[8] my passion was back.

Since the release of *Doughnut Economics* in 2017,[9] Raworth's model has been the dominant template for sustainable design. She advocates for types of growth that balance 'the needs of the living people with that of the living planet'. At the design festival, she poses some critical questions to ask when applying these models: What is our purpose? How can we best contribute? How are we networking and who are we connecting with to enable this purpose? Whose voices count in this process? Who owns these design interventions? How are these systems financed and what do they demand in return? In other words, are humanity's actions and structures enabling the goals of regenerative and redistributive design?

Raworth points out that indigenous cultures long ago figured out this balance, while Western cultures remain outliers in their worldviews on the relations between humans and nature. Where Raworth stops, I begin. I am working on bringing the voices of the Global South into this conversation. I remind the audience that what the design community proposes as sustainable strategies have been in practice for generations among indigenous communities outside the West.

Indigenous Approaches to Design

Indigenous communities across the Global South have, for centuries, related to nature as a sentient being, with its own moods, feelings, intelligence and rights. As designers shift from human-centred to green-centred design, it pays to get out of their comfort zones. Policymakers can learn from indigenous cultures to address the concurrent global crises. These learnings can aid them in rethinking relations between humans, nature and technology. These alternative models can go beyond the profit-centric ownership model and the extractive or instrumental approach to sustainable design.

The big four that can inspire designers and policymakers are the cultures of frugality, collectivity, subsistence and repair.

Figure 5.1. Green design

The *culture of frugality* in resource-constrained contexts in the Global South makes more from less. Business gurus in the West encapsulated this as 'frugal innovation' and 'frugal engineering' in the early 2000s, through an approach that focused on minimal design and economy of scale. This business model has made a comeback with endless seminars, labs, policies and projects on the subject.[10] Progressive economists push for the degrowth movement in which frugal living and conscientious consumption form the pathway to sustainable societies.

Women in the Global South nurture a *culture of collectivity* by allowing for the distribution of risk and caregiving. As ecofeminist Vandana Shiva asserts, 'We are either going to have a future where women lead the way to make peace with the Earth or we are not going to have a human future at all.'[11] The digital has appropriated this culture for crowdsourcing and crowdfunding, capitalizing on its networked effects.

Subsistence cultures, in Africa for example, have long practiced biodiversity for ecological sustainability. Now social entrepreneurs like the Food Innovation Network have moved into this territory. International NGOs like the Nature Conservancy, Conservation International, and the Wildlife Conservation Society and business coalitions such as Business for Nature have gotten on board with the Kunming-Montreal Global Biodiversity Framework's mission.

The *culture of repair* found among indigenous communities in Latin America and South Asia extends the lives of things. The right to repair legislation has only recently gained legal credence in Europe.[12]

The fact is that people in the marginalized majority innovate because their survival depends on it. There is an existential urgency to their innovation. They must think outside the box, think laterally, and think creatively, as their communities futures are contingent on the planet's future. As environmental activist

Sunderlal Bahuguna puts it, 'Ecology is the permanent economy.'[13] Despite indigenous practices, the current climate change discourse blames the 'population bomb' in Asia and Asia's growing economy as key culprits in global warming. Moreover, policymakers advocate constraining the economy for ecology. Deep-seated hypocrisies drive this mindset.

Part I: Culture of Frugality

Advocating for frugality without historical reflection and responsibility and dissociating from current consumption behaviour and context, can exacerbate existing inequalities. This well-meaning worldview may result in pitting environmental sustainability against social equality.

Decolonizing Degrowth

The Global South is often viewed as a big culprit in global warming. China and India top the charts as the world's biggest polluters outside the United States year after year.[14] *The Economist* has put the onus on Asia to change the course of environmental degradation and expressed scepticism due to what it sees as low political will, weak administrative capacity, and deep investment in maintaining the status quo.[15]

Racialized roots undergird this worldview. In the 1790s, Robert Malthus, an English cleric and economist, triggered a Western moral panic about the 'breeding' in the Global South.[16] This media narrative emerged during the industrial era when the West was experiencing high population growth and consumption. Meanwhile, the Global South was experiencing mass and early mortality due to colonialism and its violent aftermath.[17] Anxieties prevailed over these 'developing' nations taking over the world and consuming the planet's resources.

The Malthusian 'overpopulation' hysteria seems to recur with every generation. In 1967, *The New York Times* released a full-page advertisement, funded by a prominent population control group, blaming the 'Third World' for the population bomb, which supposedly 'threatens the peace of the world'.[18]

Even in 2022, the primatologist Jane Goodall appeared to have succumbed to the population bomb theory at the World Economic Forum (WEF) in Davos. She remarked that human population growth is responsible for most environmental problems.[19] This thinking is at odds with reality.

The UN refutes this persistent false narrative with cold facts—high-income countries have the highest material footprint per capita (29 metric tons per person), '60 per cent higher than the upper-middle-income countries (17 metric tons per person) and more than thirteen times the level of low-income countries (2 metric tons per person)'.[20] The world's wealthiest 10 per cent produces up to 50 per cent of the planet's consumption-based carbon emissions, while the poorest half of humanity contributes only 10 per cent.[21]

We have a problem when an American fridge uses more electricity than a typical African person.[22] Western hypocrisy runs deep. Rich countries advocate for clean energy while importing fossil fuels. They can afford to be organic, 'go local', and celebrate their air quality while dumping their e-waste in the Global South.[23] Aid agencies offer loans at debilitating interest rates for clean energy, often leaving African countries with little choice but to consume coal or look to their Asian neighbours for investment.

Blanket treatment of different contexts in the name of sustainable design can exacerbate inequality and entrench poverty. While the rich world is preoccupied with making clean energy, countries in Africa are focused on producing more energy—their pathway out of poverty.

Choice versus Condition

The degrowth movement's advocacy of 'voluntary' downscaling of consumption and production is tone-deaf to the realities of disadvantaged populations living with minimal means of survival. 'There is an important difference between frugality as a choice and frugality as a social condition,' explains Brazilian development economist Roldan Muradian.[24]

Moreover, it is false to assume an incompatibility between socieconomic mobility and environmental care. The degrowth movement fails to recognize that disadvantaged groups view dignified living and planetary well-being as symbiotic and fundamental to their place in the world. Muradian believes this movement of frugal consumption is a well-meaning but deeply delusional ecological project driven by educated, middle-income European groups.

Marginalized communities are not 'too poor to be green'.[25] Greening is not something indigenous people 'do', but is essentially how they live.[26] While indigenous peoples constitute 6 per cent of the global population, they protect 80 per cent of the world's biodiversity.[27] The Sámi people living in the Arctic, for instance, can teach us co-existence through their core principle of *eenâm lii eellim* (land is life), which shapes their everyday behaviour with nature.[28] Many indigenous communities across the world share similar worldviews in which nature enables their culture's existence. This stands in contrast to Western extractive perspectives that commodify nature to serve people.

Muradian is not alone in critiquing some of the favourite eco-friendly design ideas of our time. Labour market specialists Jana-Chin Rué Glutting and Apoorva Shankar argue that well-meaning trends that strive to achieve a zero-waste system, such as the circular economy, apply a double standard to the Global South.

Anglo-Saxon sustainability translates to high-tech solutions with little focus on the root cause of production and consumption. The circular-economy model of 'going local' for 'clean living' obfuscates the reality of globally integrated supply chains that systematically export waste from the North to the South.[29]

The data economy has added a new layer of racialized double standards to digital value chains. The metaphor 'data is the new oil' is now a material reality. Data centres are big culprits in global warming, alongside single-use plastic items, automotive pollution, and fast fashion.[30] With the rise of the next billion users, the question remains: Who foots the bill for the next billion carbon footprints? Is digital inclusion coming at the price of environmental protection? Pitting inclusion against climate justice is a no-win undertaking. With the legacy of colonial resource extraction continuing through 'data colonialism'[31] practices to build AI-based tools, we need a concerted global effort to build digitally inclusive futures alongside a sustainable, planetary one.

AI Footprints

Data centres have a bigger carbon footprint than the aviation industry.[32] Far from being abstract, invisible and intangible, data is very much physical and energy-consuming. The cloud is grounded by cables and concrete. From sensors to streaming, our everyday digital tools demand a slice of the energy pie.

Data is a critical commodity in the global AI race among governments and industries alike. Tech companies compete fiercely for what is viewed as a scarce resource. Youth in the Global South become data mines in low-rights environments where businesses can operate without the burden of the privacy regulations that exist in the West.[33] To stay competitive in the data cold war with China, US companies go elsewhere to sharpen AI, improve their cryptocurrencies, experiment on the blockchain, and develop new products and services that can be sold in wealthier markets. These exploitative actions come at an added environmental cost. Worldcoin is a case in point.

Sam Altman, best known as the former CEO of OpenAI, ChatGPT's parent company, co-founded Worldcoin to build the largest biometric identity system and financial network as a public utility, 'giving ownership to everyone'. He promises digital democracy—nothing short of radically making the world into a better place through tech. Altman envisions a postnational and postcapitalist world where all people on the planet can collectively enjoy the benefits of AI. Worldcoin uses a specially designed chrome orb that resembles a black crystal ball to collect iris scans in exchange for 'free' coins.

Collective ownership demands individual investment in the form of personal biometric data. On the ground, however, an investigative team from *MIT Technology Review* discovered jarring practices.[34] They interviewed thirty-five users, recruiters and workers in six countries—Indonesia, Kenya, Sudan, Ghana, Chile and Norway—and observed the user onboarding process at a registration event in Indonesia. The team interviewed social and tech experts, and discovered major privacy violations as Worldcoin scanned hundreds of thousands of eyes, faces and bodies in twenty-four countries, the majority of which were in the Global South. Web 3.0 experts expressed serious concerns about the security of the orb and the biometric data because of the single identity goal—the opposite of the decentralization that defines blockchain technology. To build privacy, Worldcoin broke privacy.

Crypto altruism comes at a significant carbon cost. In the last decade, 300 million people have begun using cryptocurrencies.[35] This crypto-mania is accompanied by energy-intensive Bitcoin mining that is equivalent to a small country's electricity consumption on an annual basis. With the Russian invasion of

Ukraine sending energy prices soaring across the world, many countries are tightening the rules. China's crackdown on the crypto industry has sent Asian Bitcoin nomads to seek alternative homes for their hardware, including in China's nemesis—the United States. Bitcoin miners look for predictable regulation, stable relations with governments and affordable power.[36] The search for a hospitable turf seems to be getting harder. Their start-up energy demand is equal to that needed to power downtown Dallas, making it politically contentious even in freewheeling Texas.

Bitcoin Babushkas

In the crypto-against-carbon debate, perceived value is key. Crypto enthusiasts defend themselves by arguing that they aim to dismantle traditional financial empires that have long been 'too big to fail'. Crypto is not merely about creating a new global currency.

Banking systems need a radical overhaul. In 2023 alone, cascading collapses and the crises of Silicon Valley Bank, Signature Bank and Credit Suisse remind us only too grimly that the financial system needs a new, responsible, ethical design of doing.[37]

Moreover, the WEF tells us that pitting carbon against crypto is a privileged Western worldview.[38] The WEF underlines that 80 per cent of the world's population has limited freedoms and little access to stable economic systems. 'Bitcoin babushkas',[39] grandmothers in Russia, drive mining to protect their family's savings. Young Nigerians see Bitcoin as a symbol of protest against the redesign of their currency, the naira, which became a political band-aid applied to Nigeria's economic crisis.[40]

For people in Kenya, Venezuela and Brazil, crypto is a safer bet than traditional currencies. They have little choice but to adopt crypto as they desperately seek to preserve their savings, receive remittances, purchase basic goods, and keep their businesses alive through basic fiscal transactions. This is a big reason why emerging markets continue to rank as leaders in the grassroots adoption of cryptocurrencies. According to the 2022 Global Crypto Adoption Index published by Chainalysis, setting aside the United States, Vietnam, the Philippines, India and Ukraine lead the rankings.[41]

The Global South's crypto uptake is more reflective of the failure of the state and the market than it is proof of crypto's superiority. Crypto's volatility and instability doubly punish those already suffering from very few options. We need independent global watchdogs to ensure that speculative technologies do not speculate at the expense of those most vulnerable. Promoters of innovations like Worldcoin need reminding that the process matters as much as the product and requires building regulation, governance and enforcement at a global scale.

Imagined affordances of security, accountability and trustworthy intermediaries need the boring essential bureaucracies to transform aspiration into reality. Otherwise, new technologies will offer false hope to those deprived of legitimate options. The value of energy use is as much determined by culture as it is driven by carbon. How then does one build a shared value system that puts the planet and societal equity on par for mutual flourishing?

Part II: Culture of Collectives

The power of collectives manifests through a critical mass of people exercising agency and not merely through the cumulation of people and data. It takes an imaginative re-engineering of templates, codes and labels to redefine power relations through the force of numbers.

Engineering Bio-empathy

Templates are the litmus test of our times. What we choose to keep in and leave out says something about the stories we want to share.

When you crack the code, you crack the culture embedded within it. Values dictate a variety of ways in which we can construct a story, justify a practice, and mobilize an action.

'If thinking were easy, more people would get their kicks that way.' This was the byline for Alternatives, a Bangalore-based critical-and-creative thinking centre set up in the early 1990s. I remember reading their cryptic advertisement for a workshop in the *Deccan Herald* as a teen growing up in India. It promised abstract possibilities of storytelling to change narratives. I liked not knowing what I was getting into. I was a teenager, after all. Their centre was a beautiful chaos, with posters from US activist Ralph Nadar, public intellectual Noam Chomsky, and naturalist Henry David Thoreau. It had a '60s vibe in the colours, codes and curated slogans in designer fonts scattered across rooms.

Only five of us had signed up for this Bangalore-based workshop. The organizers took us to Cubbon Park for our first exercise and told us to imagine being marooned on an island. Our new tribe would need new rules of engagement between ourselves and with the natural ecosystem around us. It was the first time I was made to account for nature in thinking of society. The group's default mindset was instrumental. We devised a plan of natural extraction, a template that became the basis for our social design.

As architect Indy Johar remarked at the Design for Planet Festival, 'Everyone and everything is entangled ... As designers we need to take responsibility for entangled values.'[42] To build templates with bio-empathy helps to recognize that we are stewards and not owners of the nature that surrounds us.

A decade later, I was hired by PlanetRead, a Pondicherry-based non-profit organization in south India. They sent me to work in Kuppam, a rural town in Andhra Pradesh, for a year. My job was to execute a World Bank-funded project on 'karaoke for social change', using same-language subtitles to sustain recently acquired literacy skills and bring about social change.[43] During that time, my team and I built a repository of folk songs for YouTube. It turned out that folk songs were not naturally suited to the audio cultures of YouTube. For one, they could go on for hours, especially at village festivals. Their instrumentals, pitches and lyrics were not in sync with the acoustics of Tollywood or Bollywood. In addition, folk content was either too religious or too political to be considered 'pop'. We designed a template to make the rural palatable to the urban, until it resembled a convenient fiction.

The Power of Labels

In a data-driven world, labelling practices organize the knowledge that shapes our thoughts and feelings. The father of UX research, Don Norman, has mainstreamed the eco-friendly design approach by advocating that designers shift from being 'unintentionally destructive to being intentionally constructive'.[44] This includes critically examining how we label people, animals and things. For instance, Norman points out that how we choose to label specific plants as 'weeds' and certain insects as 'pests' has little to do with their intrinsic value. This arbitrariness assigns plants and animals a value that dictates the amount of empathy they will receive.

This treatment can go both ways. Through relabelling, we can retell a story. In my book *The Leisure Commons*,[45] I applied the metaphor of public parks to reveal the underlying design structures and ideologies that shape the internet. I used the label of 'corporate parks' to highlight the growing privatization of the internet. I used 'fantasy parks' to emphasize the branded empires of sponsored digital spaces we play within and 'walled gardens' from the eighteenth century to capture the enclosures of public parks, drawing parallels to the closing of open web architectures.

Such entangling of histories and geographies are needed to remind us that what we value—our freedoms, play, generosity with each other—are not unique to a point in time or to a particular culture or group.

We Are All Animals

Labels of care require disrupting clusters and categories that people have come to believe are normal. Retraining datasets needs constant guidance, nurturing, and a roadmap to build and sustain empathy. *The Internet of Animals*, author Deborah Lupton is quick to point out that Western cultures have long labelled humans as separate from and more than animals, when in reality 'people are animals too, however much many of us like to forget or deny this reality'.[46] The repositioning of humans as animals may generate more empathy for animals.

By taking this perspective, consumers may experience moral outrage about the unprecedented scale of factory farming and the culling of millions of animals to safeguard human health. Recognizing the 'speciesist bias in AI'[47], how certain species are being coded with less value than others, can help build solidarity across species. Image recognition software uses commodity terms such as *hog* and *livestock* for pigs. Food terms populate image databases where fish are found as angler trophies, rather than in their natural habitat.

While these demeaning and devaluing terms can negatively influence how consumers relate to certain species, a positive spin can be just as damaging. Trained datasets portray birds in natural environments when only one-third of all living birds exist in the wild. Cows, pigs and chickens are shown in free-range environments even though a majority of them are in factory farms.[48]

Lupton conducted an online survey about animals and digital care and asked her respondents an open-ended question about the kinds of design they would like to see to help them better understand and care for animals. Many of them suggested an app to help them with their care routines for pets: reminders to walk, feed, change water, deworm, vaccinate, groom, and perform other such mundanities gained prominence as respondents envisioned their ideal care app. While some found it desirable to track their household animals' every move via smartphone, most participants wanted a new platform to help them understand animals better—software and devices to 'translate the thoughts, feelings and needs of their pets and better communicate with them'.[49]

Labels stick when they fit into a larger and more compelling story. Storytelling as a tool and technique comes with a track record of helping us relabel ourselves more humbly in the larger natural ecosystem.

Part III: Culture of Subsistence

At the core of subsistence is the value of diversity. Subsistence agriculture has long illustrated the advantages of diversifying our ecology for mutual flourishing and regeneration of soil and produce. Women have led this movement and continue to do so, decentring the paradigm of capitalism and emphasizing that of care. We must adopt this worldview if we are to recode our way out of the climate crisis.

Carbon, Code, and Culture

Software without stories is a body without a soul. Many of us feel our phones are extensions of ourselves. What stops us from feeling the same with nature? By transforming a bionic self into an eco-self, nature and technology can cohabit in worlds of our making. This transition is no easy task. It demands fresh narratives, new worldviews and alternative actions.

South African anthropologist Rosa Boswell explains how oceans shape people's identities through African storytelling, or what she calls 'blue heritage'.[50] She and I are part of the April 2023 cohort of fellows at the Rockefeller Foundation's Bellagio residency, which brings together artists, writers, activists, policymakers and scholars from diverse disciplines to work on their projects in Bellagio, Italy.

As part of its tradition, residents give evening talks about their work and share worlds of intellectual habitation. Boswell's talk on South African indigenous narratives on nature and culture came at an opportune time as I wrote this chapter. She brought entangled values to life when connecting the desert and the ocean in Namibia. These seemingly unrelated topographies are inextricably linked as southwestern winds from the ocean push fog into the desert, giving life to what would otherwise be a deeply hostile environment. The Namibian desert enjoys the status of a world heritage site as over half of the animal species found in its dunes do not exist anywhere else in the world.[51]

Stories from communities that live inland and at the ocean capture a deep understanding of the interrelatedness of things, animals and places, that can be transposed to other spaces of inhabitation, including our digital worlds.

All Flourishing Is Mutual

The Amazon is a complex material ecosystem and a metaphor for design thinking in an ecological way. Social innovation, climate justice, political storytelling and design can collectively serve a post-human perspective if we remind ourselves that 'all flourishing is mutual'.[52]

'People do not confine themselves to landscapes,' says Boswell, when speaking of constructing identities; neither does our communication with our smartphones and digital platforms. Every email we write, every song we stream, every picture we upload tells a story in space and time. Batteries that power our smart devices are entangled with the mining of minerals that go into their making and result in dead yaks in the Tibetan River.[53] Electric cars that promise a clean future are built on the dirty politics of back-breaking labour and induce drought in the so-called lithium triangle of Argentina, Bolivia and Chile.[54] Lithium mining in Chile consumes 65 per cent of the region's water, driving people and animals elsewhere. Cobalt, another critical ingredient for electric vehicles, brings cascading environmental and human damage as it demands extensive manual child labour in what are euphemistically labelled 'artisanal mines'.[55]

The computing cloud is brought down to earth in data centres and cooled by bodies of water around them. Environmental activist Teal Lehto remarks that we need to instil an 'ethic of water conservation in every industry' for the digital to be sustainable.[56] Financial geographer Dariusz Wojcik connects the dots through an 'atlas of finance'[57] drawing out the global political economy of green finance. Maps tell stories of where data centres are built, global labour is recruited, minerals are extracted, and devices are manufactured, and of where hardware goes to die. The catchphrase 'follow the money' gains renewed salience in this greening endeavour.

What Boswell does with the ocean, ecological anthropologist Eduardo Kohn does with the forest in his book *How Forests Think*. His four years of ethnography among the Runa of Ecuador's Upper Amazon reveal intimate physical and spiritual encounters. Kohn compels us to recognize that 'seeing, representing, and perhaps knowing, even thinking, are not exclusively human affairs'.[58] The United Nations High Commissioner for Refugees' strategic storytelling expert Lauren Parater cautions aid agencies to keep in mind that when constructing new designs, policies and imaginaries, they need to equally attend to the 'destruction, removal, and defuturing—or the reduction of possible futures—it creates in its wake'.[59] She advises them to not forget on what is lost when they celebrate what is gained.

Entanglements happen not just with values and geographies, but across generations as well. We inherit systems, norms and codes to live by, often by brute force. In correcting the past, we face the challenge of what needs to be kept and what needs to be discarded.

Monocultures of the Machine

Eco-feminist Vandana Shiva, in *Monocultures of the Mind*[60], accused Western development organizations of imposing their infrastructures, policies and priorities on other societies, threatening biodiversity and self-generated solutions from below. Today, we need a reminder as algorithmic structures and climate-based solutions take a similar approach. In channelling energies towards green designs, practitioners and policymakers need to go beyond the instrumental logics of top-down innovations such as the Green Revolution.

The Green Revolution was an innovative technocratic solution to world food shortage in the 1970s. This agricultural reform significantly reduced poverty by embracing modern techniques of high-yield seeds variety, chemical fertilizers, pesticides, controlled irrigation and mechanization. However, these reforms have indisputably reduced biodiversity through monocultural production.[61] Over the decades, we have seen its failures reflected in farmers' high debt levels, pesticide dependence, degraded soil quality and waterlogged deserts. Today's singular solution can become tomorrow's more complex problem.

In the data economy, we face a new monocultural challenge. In 2022, two behemoths—Meta in the United States and Reliance in India—entered a partnership to mediate the food supply chain.[62]

This corporate synergy could translate to total control over retail production and distribution by combining user data in ways that could threaten market diversification. Big projects of techno-utopian scale signal little trust in people. These consolidations solve problems of inefficiency while exacerbating problems of unfair competition, poor compensation, or even lack of cooperation with those experiencing these changes first-hand—the farmers.

Part IV: Culture of Repair

The virtue of repair is that it mends not just things but also trust. If we see our relationship with everyday things as long-term and meaningful, then we will design in ways that allow for repurposing, auditing, and fixing things in ways that can renew and regenerate. This demands a recognition that we are interconnected and live in a system of wonderful complexity, where repairing a part can change the whole.

Generation Lost

In her travels through rural China, activist-artist Xiaowei Wang speaks of a blockchain utopia entering farming communities.[63] Ledgers and decentralized records for proof of validation are messiahs in this futuristic technoscape. Even the chickens go through proofing! Wang found several AI-enabled organic chicken farms where chickens have tamper-proof ankle bracelets with unique QR codes to track their steps, weight, location and diet. As Chinese customers' trust in the market falls, these scans rise so customers can 'bond with their dinners'.[64]

AI generates a new intimacy with our food. Though rural farming was once invisible and inaccessible, today we have it in our living rooms and on our tablets and in our apps. FoodTech venture capitalist firms like Bits x Bites promise a shift in trust in food safety from the social to the technical. These impressive and futuristic-sounding efforts do little to address a problem the agricultural sector faces worldwide: few farmers want to farm any more.

While venture capital sees profit, the new rural generation sees loss. Much of what plagues this sector has little to do with tech, and more to do with the fair distribution of profit, diversified produce for regeneration, and structured resilience for farmers in terms of adequate safety nets if they are to experiment and invest in new tech for old problems.

The Sachet Lesson

The late business guru C.K. Prahalad and his 1990s 'bottom of the pyramid' model convinced companies and governments to look at the global poor as a legitimate market.[65] He argued that if we shifted our perspective to what these populations need, we could profit through economies of scale and do good at the same time.

Prahalad's inspiration came from the shampoo sachets of 1970s India. Chinni Krishnan, a farmer turned entrepreneur, realized that he could make shampoo affordable to rural consumers by selling it in small packets instead of big bottles.[66] This simple innovation sprang from observing a stark difference between middle- and lower-class consumer groups—while middle-class workers enjoyed monthly salaries, lower-class workers were daily wage labourers. This led to a radical overhaul of the Indian consumer market.

The sachet principle became a global market dictate for multinationals like Unilever and P&G to serve the Global South. It was a win-win solution for people and businesses, but not for the environment.[67] Being small, these packets fell through the cracks of recycling. They clogged drains, polluted water sources, littered the streets and ruined pristine landscapes. Landfill usage was exponentially increased by this everyday waste. Worse, it created a disposable culture among communities that historically would repair, recycle and reuse. Going against plastic is no easy feat. Plastic is durable, inexpensive, heat and water resistant, and essential to billions of marginalized groups in difficult social conditions. The core failure of the bottom of the pyramid model was that it did not take into account that things are connected, which can result in a cascading effect of solving one problem but creating another, even larger one.

Parachuting Cats

Figure 5.2. Wicked problems

Once upon a time, 14,000 live cats were parachuted into Borneo Island in Southeast Asia.[68] In the 1950s, the World Health Organization (WHO) tried to help the Dayak people of Borneo eradicate malaria in the region. Their quick-fix solution was to send helicopters to spray dichlorodiphenyltrichloroethane (DDT), an insecticide, over the region. In the short term, it worked. The mosquitoes died. Over the long term, however, it started to poison other insects and moved up the food chain, killing most of Borneo's cats. This led to an explosion of rats and bigger problems of typhus and plague. The WHO had to respond immediately, taking the desperate measure of parachuting cats into this island at an enormous cost to control the rat population. Two decades later, the United States banned the use of DDT, recognizing that it causes more harm than good.

What we have here is 'a wicked problem',[69] an issue that lacks a clear solution due to its multifaceted nature and diverse stakeholder perspectives, as well as the emergence of new and often bigger challenges during attempts to tackle an existing one. To address a wicked problem and institute sustainable change, organizations need to recognize the interrelationship of people, places and practices. We need to apply this type of thinking to some of the biggest challenges we face today, such as pandemics, climate change and data governance. The common factors across the board are the intrinsically global nature of these problems and rising inequality where the negative impacts are felt most by marginalized populations. In the case of making cyber policies fair and just, we need to consider that data is inherently social—our everyday thoughts and feelings are not just extensions of ourselves but of how we are members of a global society. When we embark on problem-solving, we need to consider the humanity of the culture of code that steers us towards a joint future.

It's Not a Contest

To ensure that we do not pit equity against sustainability, we need to invest our energies in finding solutions that address both values simultaneously. In 2022, an all-female team of students won the Earthshot Prize for coming up with an alternative to plastic packaging.[70] Using seaweed cuts gas emissions by 36 per cent, keeps profits at 41 per cent and reduces plastic use by 95 per cent. The team members came from different disciplines—law, international policy and finance—and saw plastic waste as an intrinsically global issue. Moreover, they sought to add value and not just solve a problem, as they outlined new opportunities for livelihoods for coastal communities threatened by climate change.

Repairing our current systems has limits. While the culture of repair is commendable and often necessary, it still needs to work with that which is inherited, broken, fragile and volatile. Designers, thinkers and doers must remind themselves that in the process of creating a new tool, they may be creating a new legacy of practice.

There is deep responsibility in building new templates, codes and labels that must work in tandem to ensure that both humans and the planet flourish. These become the building blocks upon which the ecosystem of future life will rest.

Epilogue

NEETA ATTENDS HER BROTHER'S ONLINE CLASSES THROUGH HIS phone. She goes to a government school while he goes to a private school in India. They and their friends google assignments for clarity. She complains that her teachers publicly berate her for asking questions, but 'Google will never insult me like this'.[1] Even parents are on board with online learning. As one remarks, 'YouTube is better than school. The clever person will learn from YouTube. They will take admission in a government school and then learn from YouTube.'[2] Teachers also throw their hats into the ring. Kamlesh, a government-school teacher, has set up his own Instagram account, where he experiments with teaching methods through Reels while earning a little cash on the side.

This is a snapshot from a digital learning project that FemLab undertook, beginning in 2022.[3] We focused on resource-constrained communities in India, which are the most vulnerable when it comes to losing out on access to education. We found that youth have high aspirations—they want to become doctors, engineers, entrepreneurs. They want a fighting chance in a world structured by injustice, but changing an educational system, a social norm and a patriarchal culture takes time.

Using digital tools is a quick way to make a dent, carve a niche, and etch a space in a relatively safe way. But while YouTube saves the day, schools are here to stay. School is synonymous with education, especially for parents with limited economic means, who view this formal pathway as the single most powerful way for their children to escape poverty. Both learning spaces are fraught with harm and challenges, but signal progress for the next generation. Digital technology, much like school, is synonymous with hope.

When we turn to the West, the education system, much like other social systems, appears deeply fragile. New AI technologies like ChatGPT could disrupt traditional ways of learning, teaching and assessment. Tech monopolies pose an existential threat to the social systems people have come to cherish. Resentment leads to resignation as users inhabit digital spaces they loathe, consume content they condemn, and engage with tools they view as toxic to their lives. This book makes the case that this pessimism bubble, while persistent, represents a small fraction of the world's perspective. Even as despair drives design, policy and public sentiment in the West, the rest of the world remains contagiously hopeful.

Contagion of Hope

With close to 90 per cent of young people worldwide residing in the Global South, hope is the global norm.[4] These youths yearn to disrupt and dismantle systems that continue to fail them. They are forward-looking and optimistic. Digital platforms, despite the risks and harms, are often more pliable than existing institutions and cultural norms. Afghan girls and women, deprived by the Taliban of access to education, parks, markets, and even salons, cope with this medieval treatment by finding sanity online.[5] They watch

YouTube tutorials and Instagram fashion Reels and share jokes with strangers from the isolation of their homes.

Millions of young resellers in Bangladesh use Facebook to do business, especially when the COVID-19 pandemic shut down their markets. Some imams in Saudi Arabia with more liberal values experiment on X, making piety accessible to their young followers. As nuance paves the way for negotiation, Global South users have little choice but to hack creativity, circumvent state control, and build proxies for self-expression. The new algorithmic cultures promise new ways to make a living, escape loneliness and collectively champion a cause.

The binary of good and evil is fitting for children's stories but not for reality. As we grow up in digital worlds that are intrinsically cross-cultural, we need to ask questions that mirror that ethos. Tech bros should resist claiming that new tech will replace entire institutions and cultures; civic actors need to shelve interrogations that serve as a litmus test for which side people are on; media pundits need to resist queries of causality, such as, will this new tech be a kill switch for democracy/romance/creativity/community/civility? Instead, they should ask questions about why people and institutions do what they do and how we can steer them to optimize technologies for human and planetary well-being.

In the fast-paced world of digital innovation, insights on AI futures may appear slow and quickly outdated. Trends rise and fall and threaten to take these visions with them into the shadows of oblivion. In the graveyard of apps, we find punditry buried. This angst, however, applies to those who insist on believing that novelty in tech translates to novelty in social life. Human nature, on the contrary, is predictable in its response to new technologies. Society tends to react to new technologies with utopian and dystopian futuristic visions, and moral panics. That leads us to reactively approach designs and policies. The fact is that all innovations are intrinsically humane, complex, contradictory, and often become mundane over time. Novelty becomes fringe. New tools and techniques, if they are lucky to survive, become embedded in the fabric of our everyday lives—as ideas, cultures, and perhaps even as institutions—making way for the next wave of novelty. This should serve as a comfort to readers who seek to move from pessimism to possibility.

The Future of AI Is Manual, Material and Multicultural

Mainstream media headlines on the rise of AI declare the end is near. An article in *The Guardian* states, 'The Future of AI Will Fill You with Unholy Terror'.[6] *The Nation* declares 'The Future of AI Is War'[7] while the BBC warns us 'Artificial Intelligence Could Lead to Extinction'.[8] Even former OpenAI CEO Sam Altman echoed this fear in his testimony to the United States Congress in which he pushed for state regulation to ensure that the benefits outweigh the harms to society.[9] If readers want to ground themselves to cope with this existential threat, there are few spaces of solace. Every act you do and every word you say is up for automation and erasure. In this climate, this book, true to its optimistic commitment, promises a future with AI that is rooted in the manual, the material and the multicultural.

The Manual

As the world grapples with the loneliness epidemic and people yearn for real connection, it is worth imbibing indigenous and feminist wisdom that values the manual as vital to being humane, beyond the digital, the market and the state. Over the internet, a hug can only go so far. Human touch builds intimacy. Face-to-face interaction nurtures empathy and physical presence stirs hope and belonging; we are social creatures, after all. There is something to be said for the joy of shared experience.

People who fix a car, mend a fence, build a shed, knit a sweater, or cook a meal increasingly evoke respect. The AI age parallels one in which societies are starting to re-value repair culture, organic living, going local and being offline. However, existing systems are designed to valorize the head over the hand. Our institutions are geared towards deskilling youth in their everyday lives while upskilling them for some corporate niche. The manual offers an opportunity to slow down time, to savour the present, and to immerse ourselves in the pleasure of collaboration. It offers intrinsic value and personal satisfaction in being autonomous yet useful to others. The manual has a profound role to play in satisfying this fundamental human tactile need.

When handiwork is valued, there are double standards. Those who are good with their hands in the Global North are celebrated for their authenticity, individuality and creativity. Those in the Global South, however, are viewed as mass producers, cheapening the allure of the manual. Moreover, what counts as automation can be a market fabrication. The AI value chain for a business[10]—all the phases in its business cycle from the creation of a digital product or service to the delivery to market and everything in between—needs reassessment. The fact is that most of this AI cycle is fuelled by the sweat of people in the Global South: the physical labour of data feeders, labellers, content moderators, coders, auditors, data stewards, and those working in renewables and grid maintenance, all striving to keep the AI wheels turning.

Western corporations and policymakers need to acknowledge the full life cycle of the future of AI, much of it hidden and invisible to consumers and citizens. Culture will never be completely computable, nor should it be. Intermediaries of the physical and emotive kind are here to stay. AI labourers should be integrated into their imagination, decision-making and value systems. We can learn from feminist struggles around the world that continue to resist patriarchal systems that have made women's networks of care invisible and disparaged while depending on them as essential safety nets. Good leadership in the AI era comes from those who are driven by compassion and camaraderie, recognizing and instituting feminist values when it comes to innovations.

The manual today gets a bad rap. It needs a renaissance, especially in the AI age. Its Latin root is from the term *manualis*, meaning 'of or belonging to the hand'. In antiquity, it was associated with craftsmanship, artistry and human skill. There was a certain honesty that came with handiwork. Its status changed during the Industrial Revolution as it transformed into physical acts of monotony and repetition primed to be replaced by the machine. This historical shift marked a growing devaluation of the manual, equating it with the menial.[11] Elites used the new techniques of the day to build arbitrary and artificial differentials of value between the mental and the physical, the intellectual and the manual, to systematize inequality.

Fast-forward to the present, when advanced AI tools have created an existential crisis of what it means to be human. As this book illustrates, this panic builds on the continuing legacies of patriarchy, paternalism and narratives of Western progress that negate the culture of care that people around the world crave. While this mindset has moved into the digital era, we have much to learn from long-dismissed indigenous cultures that have developed ways to imbue the manual with spiritual significance, dignity and mindfulness.

The Material

AI is the air we breathe, the water we drink, the land we stand on. It manifests in underwater cables, data centres and solar panels. The hunger of the cloud is fed by our planet. Once corporate leaders and policymakers get into that default mode of thinking of the virtual as physical, and institutionalize it in everyday decision-making, it will get easier to translate sustainable values into design. For this approach to succeed, it will be imperative to honestly assess the material costs associated with automating human needs and channel creative energies to address such challenges. As corporate leaders weigh the actual price of AI-driven automation through this ecological approach, they may be able to break out of the age-old logic: 'if we build it, they will come'. Strategic friction is needed to ensure that AI futures are approached in ways that nurture our digital, social and environmental health.

In 2023, new facilities operated by Africa Data Centres will occupy 125 acres of land in Ghana.[12] This is the tip of the iceberg. From Nairobi to Cape Town, Africa is witnessing a boom in data centres in the name of data localization, digital sovereignty and local leadership. There is a corresponding increase in water consumption to cool these data centres. As global digital communication becomes more audio-visual and the demand for unstructured data increases for advanced computing, water and data become two sides of the same coin. The United Nations has declared that by 2025, 50 per cent of the world's population is projected to live in water-stressed areas,[13] making the environmental cost of data growth a priority, lest the war on data become a war on water.

Indigenous communities have long understood the delicate balance between humans and the natural world. Their intimate relationships with the environment have given rise to profound respect for all living beings based on the recognition that every action has material consequences. Applying this wisdom to the realm of AI may generate innovative ways to assess and balance our needs for being digitally connected and having a dignified living with adequate resources. Indigenous cultures can teach decision-makers the critical value of the interconnectedness of all life forms, including humans and the land. This holistic approach offers a stark contrast to the modern era's anthropocentric view, which tends to overlook the intricate web of relationships that sustains life. By understanding that AI's material impact extends beyond human society, we can foresee its implications for ecosystems, biodiversity, and even climate change, encouraging more responsible development and deployment.

Additionally, indigenous cultures carry deep understandings of sustainability and long-term thinking. Their traditions often incorporate practices that ensure resources are used wisely, preserving the earth's vitality for future generations. This ancient knowledge challenges our fast-paced, consumer-driven approach to AI development, where the race for technological supremacy sometimes overshadows the importance of ethical considerations and environmental stewardship. Moreover, indigenous cultures have demonstrated remarkable resilience in the face of adversity, preserving their traditions and knowledge over countless generations. As leaders navigate the uncharted waters of AI, indigenous people's stories of survival and adaptation should serve as inspiration to foster hope. These narratives stand as helpful reminders that humanity possesses the capability to find solutions, adapt, and learn from the challenges that AI may bring, ultimately leading to a more optimistic future.

The Multicultural

The rise of the next billion users from the Global South promises to radically diversify our digital cultures. While tech corporations struggle to move beyond the commodification mindset, and the state and aid agencies work to shift away from their need for totalizing control, users are at work shaping the nature of global datasets and algorithmic cultures. Content creators from rural areas in Zambia and Brazil churn out multimedia content daily, playing with stereotypical scripts of poverty porn, victimhood, and racialized and sexualized selves, often through satire, humour or just mundane snippets of everyday dignified living. AI is a numbers game and the numbers work in their favour.

As this book has revealed, these youth, or what I call the 'marginalized' majority, are not waiting for policies, designs or strategies to include them. They are forging ahead by gaming the metrics, finding their niche, and hunting for cross-cultural emotive resonance that can make their content go viral. However, the responsibility to diversify datasets should not rest on their shoulders. This trend offers an opportunity for corporations, entrepreneurs and designers to embrace a multicultural and multilingual approach to debiasing and building new datasets. Diversity by default can be the aspirational rule of the day.

There is much work to do to curate multicultural datasets. When we go online to search for images of people, things and cultures outside of the WEIRD (Western, educated, industrialized, rich and democratic) context, we often face data absences or stereotypical images. For instance, most indigenous communities in Latin America are not mapped, captured or visualized. An image search for 'women at work in India' returns a stark lack of women doing household work. This situation is further exacerbated by the limited types of visuals in stock images and under Creative Commons licenses. This case can be made for audio diversity as well. It is understood today that the quality of trained datasets influences what we experience when online; what we hear, see and feel are shaped by the biases in these systems. These limited options influence how the West imagines and approaches the rest of the world. With algorithmic models being only as good as the quality of the training data, these myopic framings will amplify stereotypical views of the majority world.

Generative AI tools have the potential to create new semi-synthetic datasets in partnership with Global South creators and communities to tackle this persistent sensory and cultural bias. Addressing data scarcity and distortion has positive implications for educators, activists, entrepreneurs and policymakers driven to build understanding and empathy for diverse cultural heritages and contexts without being trapped in the given boxes of cultural and creative representation. Questions abound regarding the cultural identity, digital creative ownership and fair value, scalability, and quality of this creative content, that demand an interdisciplinary social and computational undertaking.

A multicultural dataset will enable AI algorithms to learn from a wide range of voices, providing a more comprehensive understanding of the human condition. With diverse datasets, algorithms are less likely to perpetuate stereotypes or discriminatory patterns, as they will have been exposed to a variety of perspectives and cultural nuances. As a result, AI systems will make more informed and empathetic decisions, leading to greater fairness and social justice in areas like hiring, loan approvals and criminal justice, among others. A multicultural approach to building datasets serves as a powerful catalyst in our journey towards a more compassionate society.

Decolonial Design

Decolonizing is already a buzzword. The term's popular usage in media today denotes a widespread awareness of how social injustice is inscribed in our AI worlds and there is momentum to make AI more inclusive. The challenge remains to translate this drive into design interventions.

Build Bandwidth

Reflexive fatigue is the first obstacle. Many institutions have embraced moral agendas, but few have inscribed them in their organizational workings so that people have the bandwidth to pause the business-as-usual approach and make meaningful changes to their practices. Organizations treat the decolonizing agenda as an add-on. Doing things differently may result in emotional exhaustion and burnout, which may translate to resentment towards the very groups targeted to be served. There is a reason why cooperative design efforts often result only in product testing experiments or market sentiment pilots. Tight and short-term deadlines, restrictive outcomes as measures of success, limited incentives, and the lack of a common vocabulary to understand grievances and aspirations hold people back from plunging into this rich reality.

Invest in Relationships without Agendas

As organizations grow, they can't help but get trapped in their cultural myopia. Relationship-building in public and private enterprises is typically instrumental. The term *networking* smells like work. To do things differently, and more equitably, corporate and policy leaders should have safe spaces to build ties with civic agencies, artists and community members where their organizations can organically learn from one another and chalk out common ground. This takes time, energy and long-term commitment. Every organization has internal disruptors who are able and willing to shift their work cultures from within. To identify, collaborate and deepen these relationships, we need time and space, and most importantly, an open agenda.

Get Past the 'Gotcha' Mentality

Innovation comes with failure. Bad designs, policies and code are made all the time. There is no such thing as a finished product. When products are released into the wild, they take on lives of their own. Yet, the responsibility for the impact these products have on society—especially towards those most vulnerable—still lies with the organization. Accountability should not arise only when mistakes are made. Auditors are there to do more than ensure standards are met—they serve as educators, mediators and catalysts for change by aggregating insights of the most affected groups and sharing these understandings in ways that can be incorporated into further product development.

We can get lofty about decolonial design, but at the end of the day, systemic change happens incrementally, consciously and thoughtfully, by embracing the 'boringness' of institutional reform.

Pop the Pessimism Bubble

I was in New Orleans to deliver the closing keynote at an international conference on human-computer interaction in 2022. It was my first post-pandemic, in-person appearance at a large international event, as it was for many of its more than 2,000 attendees. When I usually talk in front of a big gathering, I love to get the pulse of the room. I asked the audience to make some noise if my statements and questions resonate with them. At the beginning of this keynote, I ask how many in the crowd work in UX. Some people clap. I followed up by asking who worked in or with people in the Global South. A few people made some noise. I then asked if anyone felt like an imposter in their area of expertise. The room exploded.

***

When even tech people don't feel they belong, we have a cultural crisis at hand. Utility companies and automobile industries now compete with the data industry. Super apps push gaming, banking and romance industries into an arranged marriage. Banks compete with WhatsApp and TikTok to issue digital currencies. AI has seeped into the everyday operations of systems, shifting culture from the inside out. Academics have become influencers. Influencers have become politicians. Politicians have become entertainers online. The world feels upside down. Virtual hate leads to bodily harm, even death. Land gets occupied by the cloud.

Imposter syndrome doesn't have to be a malady; it can be an asset as it allows us to stay at the cusp of change, ready to take on the future. No discipline, expertise, or industry practice should be sacrosanct. Embracing the imposter in ourselves shows that we do not need to belong to some sector or field to feel part of a whole.

These AI trends don't have to be disorienting. If leaders truly imbibe the lesson of interconnectedness, they will start by looking at any product or service, from toothpaste to health insurance, as a cultural artefact shaped by larger social and ecological forces. They will pursue the value chain of manual labour that goes into their business, its material impact to gauge the true value, and how different cultures perceive their products and services.

In 2022, I was headhunted to serve as a director in AI impact for a Western multinational furniture company. It puzzled me at first, as I did not see the link between my CV and selling chairs. After my conversation with their team, it became clear. Their business model was pivoting to the next billion users, who are their future customers. AI is not the only black box that needs unpacking. The Global South is just as alien and incomprehensible to many businesses. What makes people do what they do is at the heart of any operation. Furniture, for example, is an aspiration that comes alive in Pinterest posts and Netflix shows. Homes are social spaces of memory and intimacy, and a chair tells the story of pregnancy and loss. Our situated contexts, our relations, our everyday institutions and our cultures shape how and what we value.

At the end of the day, if an organization wants to get on board to build hope through their AI-enabled designs and policies, they need to buy into the fact that culture, local and global, truly matters.

Acknowledgements

WRITING A BOOK DEMANDS COURAGE. AUTHORS NEED TO HAVE something to say, and that usually comes at the price of contesting popular claims, pushing against cherished beliefs and being that somebody who sounds the alarm on what needs to be untold and retold. No wonder authors never do this alone. We rely on people and institutions to give us the strength to proceed.

I was fortunate to have been accepted to the Rockefeller Foundation's Bellagio Center Residency Program in April 2023, where I got to write more than half of this book. It does help that Bellagio happens to be one of the most beautiful places in the world. I had amazing food, the company of other fellows and, most importantly, a lot of uninterrupted time to think and write. I wrote this book in the gorgeous little office called Veduta, previously occupied by legends like Maya Angelou and, more recently, Amanda Gorman. I got to run my ideas past several visiting workshop participants from different disciplines and contexts, which further sharpened my thinking.

I was doubly lucky to have had multiple experiences at this Center that further shaped this book. The first was in August 2020, when the Rockefeller Foundation invited me to be part of a consortium of technologists, philosophers, economists, lawyers, artists and philanthropists to explore how AI can be harnessed for creating a better future for humanity. In 2023, I got to continue this discussion through an event organized by Marietje Schaake and Francis Fukuyama at Stanford University, on building synergies to identify education, research and policy needs with the onset of emerging technologies in global contexts.

This book, however, was born in the field. I am not one of those Indiana Jones anthropologists who embarks on a solo mission to investigate some exotic tribe. I strongly believe that contemporary and digital anthropology is a collective effort to sustain relationships with people and institutions. This helps to give value to our participants as much as they give value to researchers like me to help think of new ways of being and doing.

My book's arguments are brought to life with storytelling from people and organizations in India, Bangladesh and Brazil, among others. I am lucky to have Usha Raman by my side, who co-founded FemLab—a feminist futures of work initiative in 2020. FemLab has now become a platform for several of my projects that I have drawn inspiration from for this book. Our goal is to centre the voices at the margins, to build digital campaigns to translate research into accessible insights for our targeted groups, and to shape inclusive technologies and designs. We have built something beautiful, where our global team continues to engage empathetically with one another. FemLab started small, and now has grown to become a collective of junior and senior scholars and practitioners across the world—and this has enabled this book in many ways.

I am grateful for seed funding from the International Development Research Centre, a Canadian grant agency, and other funding support from the UN Refugee Agency and a major design company for supporting my field projects. While I have engaged with different teams for different projects, two persons have become indispensable to me—Siddhi Gupta and Shruti Singhal. I would like to thank Siddhi Gupta, a brilliant designer and educator, for her creative illustrations in this book and for being my go-to person to help build my research teams in India, and for working closely with me to run them. I am thrilled to have Shruti Singhal as an anchor, who wears numerous hats, including that of being the backbone of FemLab's media outreach and engagements, and for her meticulous editorial support for this book.

I would like to thank the IDRC research team—Sai Amulya Komarraju, Chinar Mehta, Pallavi Bansal, Shweta Mahendra Chandrashekhar and Upasana Bhattacharjee, who have done a remarkable job pursuing organizations and women workers in the field, despite numerous obstacles, including COVID-19–related challenges. This book draws from this project, especially for the 'Feminist Futures of Work' section in the 'Surveillance of Care' chapter. I am grateful to Justice Adda, a legal and design non-profit in Delhi, and FemLab's research partner, for inspiring me to tell good stories to institute change.

I am thankful to John Warnes, Erika Perez-Iglesias and their team at the UNHCR Innovation Services for partnering on a Digital Leisure project in refugee settlements in Brazil in 2021, which was based on my book *The Next Billion Users*. I would like to thank Amanda Alencar, Daniela Jaramillo-Dent and Julia Camargo for their fieldwork collaboration, and Paula Wittenburg for her research assistance. I use some insights that emerged from this particular project in the 'Introduction' and elsewhere in this book. Furthermore, I am grateful to Natasha Rusch for her leadership in the AI-enabled anti-poaching software project in Africa, and the support from Konrad Adenauer Stiftung and the Digital Asia Hub for helping us share our work in their outlets. I used some insights from this project in the first chapter of this book.

The 'Hacking Creativity' section in the second chapter draws from the Digital Creativity project funded by a global design company. They hired me to explore how the next billion users take to digital tools to shape their everyday creativity. I am grateful to Laura Herman, a UX industry leader, who worked closely as a liaison to ensure we translate academic research into industry-related insights. She is one of those rare internal disruptors with agency that we need more of, to make headway into such synergies. I am indebted to our research leads, including Sai Amulya Komarraju, Kiran Bhatia, Siddhi Gupta and Devina Sarwatay, and research assistants Stuti Dalal, Gargi Shivanand, Manisha Madapathi and Upasana Bhattacharjee for making this project a success.

A big thank you to influencer Trupti Shirodkar for partnering with us to launch the Hacking Creativity masterclasses. Manisha Pathak-Shelat at the Mudra Institute of Communications, Ahmedabad, provided invaluable advice in the digital learning segment and continued institutional support for multiple FemLab projects. Moreover, along with UX researcher Kiran Bhatia, the three of us have collaborated several times in different research projects, which have been influential in my thinking for this book.

What has been particularly exciting for me is bringing field insights to the table when in discussion with folks in the public and the private sector who are concerned about the future of AI. I have been invited to give talks and workshops at city festivals, corporate leadership events, government panels, tech workshops and fellow events around the world, all of which contributed to shaping my thinking. I cannot possibly do justice to thank the full range of actors I have engaged with.

I recognize how rare an opportunity it is to penetrate the walls of corporate and state leadership, especially for academics. For this,

I am grateful to my speaker agent Deborah Rey-Burns for helping me gain access, and event guru Monique van Dusseldorp for her ongoing recommendations to diverse and fascinating events. I am also thankful for my friendships with CEO Shaun Wiggins at Soteryx, Lance Shields at Adobe, Charles Hayes from IDEO, Mark West at UNESCO, and folks at RNW Media, Prosus, Google, Spotify, Liberty Latin America Telecom, NetHope, UNDP, UNHCR, Europol and many others who have opened doors for me and shown receptivity towards engaging thoughtfully and responsibly in the Global South.

Academic solidarities have come from several scholars such as Nathaniel Persily at Stanford University, Mark Deuze at University of Amsterdam, Andrea Medrado at University of Westminster, Simeon Yates at the University of Liverpool, Carlos Carlos (ChaTo) Castillo at Universitat Pompeu Fabra, and many others. These networks of care have been essential for my mental health in a climate where academia can be viciously competitive and self-consuming. I would also like to thank my MIT Press editor Noah Springer for his responsive and thoughtful approach during the fragile period of writing. He gave me much-needed encouragement, along with constructive feedback, which helped enhance this text considerably. A big thank you to my HarperCollins editor Sachin Sharma, the editorial teams of both presses and the anonymous reviewers for their encouraging comments and useful advice on how to strengthen this book. I truly respect their time and energy in this book, especially at a time when everyone is stretched thin.

Lastly, when I speak about courage, my most formidable task was to ensure that I do justice to the courage of all the Global South participants that my teams and I have worked with. Despite their uphill battle due to chronic sociopolitical and economic constraints, they exude hope for a better future. The least I can do is sustain that momentum.

Bibliography

'About Us.' *Rest of World*. https://restofworld.org/about/.

Acey, Camille E., Siko Bouterse, Sucheta Ghoshal, Amanda Menking, Anasuya Sengupta, and Adele G. Vrana. 'Decolonizing the Internet by Decolonizing Ourselves: Challenging Epistemic Injustice through Feminist Practice.' *Global Perspectives* 2, no. 1 (2021): 1–8.

Adegoke, Yinka. 'Why an Abu Dhabi-Based Streaming App Is the Future of the Global Music Industry.' *Rest of World*, 2 September 2021. https://restofworld.org/2021/why-an-abu-dhabi-based-streaming-app-is-the-future-of-the-global-music-industry/.

'Africa Will Remain Poor Unless It Uses More Energy.' *The Economist*, 3 November 2022. https://www.economist.com/middle-east-and-africa/2022/11/03/africa-will-remain-poor-unless-it-uses-more-energy.

AIA. 'Materials Pledge.' American Institute of Architects. Accessed 14 May 2023. https://www.aia.org/pages/6351155-materials-pledge.

'AI for the Global Goals.' Google. Accessed 14 May 2023. https://globalgoals.withgoogle.com/globalgoals.

Alberro, Heather. 'Why We Should Be Wary of Blaming 'Overpopulation' for the Climate Crisis.' *The Conversation*, January 28, 2020. https://theconversation.com/why-we-should-be-wary-of-blaming-overpopulation-for-the-climate-crisis-130709.

Amadeo, Kimberly. 'Why Trickle-Down Economics Works in Theory but Not in Fact: The Shortcomings of Supply-Side Economics.' *The Balance*, 30 December 2021. https://www.thebalancemoney.com/trickle-down-economics-theory-effect-does-it-work-3305572.

Appadurai, Arjun. 'The Haptic and the Phatic in the Era of Globalization.' *Glocalism: Journal of Culture, Politics and Innovation*, no. 1 (2022): 1–13. doi: 10.12893/gjcpi.2022.1.2.

Ariely, Dan. 'Column: You Are What You Measure.' *Harvard Business Review*, June 2010. https://hbr.org/2010/06/column-you-are-what-you-measure.

Armstrong, Paul. '10,000 People in Copenhagen Are about to Determine a Better Future for You.' *Forbes*, 23August 2019. https://www.forbes.com/sites/paularmstrongtech/2019/08/23/10000-people-in-copenhagen-are-about-to-determine-a-better-future-for-you/?sh=6ea7e42b6aeb.

Arora, Arun, Daniel Glaser, Aimee Kim, Philipp Kluge, Sajal Kohli, and Natalya Sak. 'It's Showtime! How Live Commerce Is Transforming the Shopping Experience.' McKinsey Digital, 21 July 2021. https://www.mckinsey.com/capabilities/mckinsey-digital/our-insights/its-showtime-how-live-commerce-is-transforming-the-shopping-experience.

Arora, Payal. 'AI Isn't Going to Save Us.' *Rest of World*, 4 March 2021. https://restofworld.org/2021/ai-isnt-going-to-save-us/.

Arora, Payal. 'The Biggest Myths about the Next Billion Internet Users.' *Quartz*, 5 November 2019. https://qz.com/1669754/tech-companies-misunderstand-the-next-billion-internet-users.

Arora, Payal. 'Bottom of the Data Pyramid: Big Data and the Global South.' *International Journal of Communication* 10 (2016): 1681–1699.

Arora, Payal. 'The Folksong Jukebox: Singing Along for Social Change in Rural India.' *Asian Journal of Communication* 22, no. 4 (2012): 337–352.

Arora, Payal. 'General Data Protection Regulation—a Global Standard? Privacy Futures, Digital Activism, and Surveillance Cultures in the Global South.' *Surveillance & Society* 17, no. 5 (2019): 717–725.

Arora, Payal. *The Leisure Commons: A Spatial History of Web 2.0*. New York: Routledge, 2014.

Arora, Payal. *The Next Billion Users: Digital Life beyond the West*. Cambridge, MA: Harvard University Press, 2019.

Arora, Payal. 'Who Is in Charge of the Future of the Internet?' Tedx Talks, Erasmus University, Rotterdam, 10 January 2017. https://www.youtube.com/watch?v=kiQa78tgbMk.

Arora, Payal, Evelyn Austin, and Geert Lovink. 'Stuck on the Platform Reclaiming the Internet.' Panel at V2_Lab for the Unstable Media, 9 November 2022. https://v2.nl/events/stuck-on-the-platform-reclaiming-the-internet.

Arora, Payal, Usha Raman, and René König, eds. *Feminist Futures of Work: Reimagining Labour in the Digital Economy*. Amsterdam: Amsterdam University Press, 2023.

Bajaj, Tejika. 'Bengaluru Based Startup FRND Raises $6.5 Million in Series A Round of Funding.' Startup Story, 21 December 2021. https://startupstorymedia.com/insights-12-21-frnd-raises-6-5-million-in-series-a-round-of-funding/.

Banerjee, Sumanta. 'Towards a Critique of Language Imperialism.' *Economic and Political Weekly* 27, no. 28 (1992): 1473–1475. https://www.jstor.org/stable/4398617#metadata_info_tab_contents.

Barnwal, Pallavi. "Indians Don't Talk about Sex—So I Help Them." *BBC News*, July 11, 2021. https://www.bbc.com/news/stories-56838660.

Benner, Chris, and Manuel Pastor. *Solidarity Economics: Why Mutuality and Movements Matter*. New York: John Wiley & Sons, 2021.

Bennett, Michael Timothy, and Yoshihiro Maruyama. 'Intensional Artificial Intelligence: From Symbol Emergence to Explainable and Empathetic AI.' ArXiv, preprint submitted 23 April 2021. https://arxiv.org/abs/2104.11573.

Berendt, Bettina. 'AI for the Common Good?! Pitfalls, Challenges, and Ethics Pen-Testing.' *Paladyn, Journal of Behavioral Robotics* 10, no. 1 (2019): 44–65.

Bhatia, Kiran Vinod, Payal Arora, and Manisha Pathak-Shelat. 'Good Girls Don't Go Online: Unpacking the Quotidian Playful Resilience Influencing Girls' Social and Digital Engagements.' *International Journal of Communication* 15 (2021): 19.

Bhatia, Kiran Vinod, Manisha Pathak-Shelat, and Payal Arora. 'DIY Education in the Digital Era: Youth-Driven Learning Strategies and Curricula for the Future of Work Opportunities.' *Education and Information Technologies* (2023): 1–20.

Bhattacharjee, Upasana, Payal Arora, and Usha Raman. 'Will E-commerce Revive Bangladesh's Artisanal Economy?' *GlobalDev Blog*, November 8, 2021. https://globaldev.blog/blog/will-e-commerce-revive-bangladesh%E2%80%99s-artisanal-economy.

Borau, Sylvie, Tobias Otterbring, Sandra Laporte, and Samuel Fosso Wamba. 'The Most Human Bot: Female Gendering Increases Humanness Perceptions of Bots and Acceptance of AI.' *Psychology & Marketing* 38, no. 7 (2021): 1052–1068.

Boswell, Rosabelle, David O'Kane, and Jeremy Hills. *The Palgrave Handbook of Blue Heritage*. London: Springer Nature, 2022.

Brackett, Marc, and Christina Cipriano. 'Emotional Intelligence Comes of Age.' *Cerebrum* (2020). https://www.ncbi.nlm.nih.gov/pmc/articles/PMC7409775/.

'Bringing Light to the Dark Side of TikTok's Algorithm.' Within Health, updated January 9, 2023. https://withinhealth.com/explore/tiktok-research.

Broadbent, Emma, John Gougoulis, Nocole Lui, Vikas Pota, and Jonathan Simons. 'What the World's Young People Think and Feel: Generation Z: Global Citizenship Survey.' Varkey Foundation, January 2017. https://www.varkeyfoundation.org/media/4487/global-young-people-report-single-pages-new.pdf.

Brown, Lee. 'Iranian Couple Sentenced a Decade in Prison for Dancing in the Street.' *New York Post*, 31 January 2023. https://nypost.com/2023/01/31/iranian-couple-get-10-5-years-prison-for-dancing-in-street/.

'Building for Everyone Everywhere.' *Next Billion Users* (blog), Google. Accessed May 14, 2023. https://nextbillionusers.google/.

Burma, Roshnee. 'World Sexual Health Day 2020.' *Are We Listening?* (blog), International Planned Parenthood Federation. 4 September 2020. https://sar.ippf.org/blogs/are-we-listening.

Burton-Bradley, Robert. 'China's LGBT Community Caught Up in Xi Jinping's Widening Crackdowns on Big Tech, Education and Celebrities.' *South China Morning Post*, 7 October 2021. https://www.scmp.com/news/people-culture/gender-diversity/article/3151504/chinas-lgbt-community-caught-xi-jinpings.

Buzz Staff. 'Kili Paul Honoured by Indian High Commission in Tanzania for Winning 'Millions of Hearts.'' *News18*, 22 February 2022. https://www.news18.com/news/buzz/kili-paul-honoured-by-indian-high-commission-in-tanzania-for-winning-millions-of-hearts-4798532.html.

Carrick, Chelsea. "Our Main Demand Is Not to Get Killed': Mexican Women Find Safety in Location-Tracking Facebook Groups.' *Rest of World*, 12 September 2022. https://restofworld.org/2022/mexico-women-location-tracking-facebook-whatsapp-groups/.

Ceuterick, Maud. 'An Affirmative Look at a Domesticity in Crisis: Women, Humour and Domestic Labour during the COVID-19 Pandemic.' *Feminist Media Studies* 20, no. 6 (2020): 896–901.

Chandrashekhar, Shweta Mahendra. 'Migrant Workers and Digital Inclusion in the Construction Sector in India.' In *Feminist Futures of Work: Reimagining Labour in the Digital Economy*, edited by Payal Arora, Usha Raman, and René König, 167–184. Amsterdam: Amsterdam University Press, 2023. https://doi.org/10.2307/jj.2711713.17.

Cheikosman, Evin. 'Why the Debate about Crypto's Energy Consumption Is Flawed.' World Economic Forum, 23 March 2022. https://www.weforum.org/agenda/2022/03/crypto-energy-consumption/.

'China Divided as WeChat Deletes LGBT Accounts from Platform.' *BBC News*, July 8, 2021. https://www.bbc.com/news/world-asia-china-57759480.

Cizek, Katerina, and William Uricchio. *Collective Wisdom: Co-creating Media for Equity and Justice*. Cambridge, MA: MIT Press, 2022.

Clemente, Drew. 'Why Do Designers Hate Canva?' *Website Builder Insider*, 27 September 2022. https://www.websitebuilderinsider.com/why-do-designers-hate-canva/.

'Climate Change Mitigation.' United Kingdom Green Building Council, accessed 13 December 2023. https://ukgbc.org/our-work/climate-change-mitigation/.

'Climate Justice.' Sustainable Development Goals, United Nations. Last modified May 31, 2019. https://www.un.org/sustainabledevelopment/blog/2019/05/climate-justice/.

Climate Tech Report 2022. London: TechNation, 2023. https://technation.io/report/climate-tech-report-2022/.

'The Cloud: Is It Bad for the Environment?' *BBC News*, 19 February 2023. https://www.bbc.co.uk/newsround/64625655.

Costanza-Chock, Sasha. *Design Justice: Community-Led Practices to Build the Worlds We Need*. Cambridge, MA: MIT Press, 2020.

Couldry, Nick, and Ulises A. Mejias. 'Data Colonialism: Rethinking Big Data's Relation to the Contemporary Subject.' *Television & New Media* 20, no. 4 (2019): 336–349.

Crawford, Kate. *The Atlas of AI: Power, Politics, and the Planetary Costs of Artificial Intelligence*. New Haven, CT: Yale University Press, 2021.

Dados, Nour, and Raewyn Connell. 'The Global South.' *Contexts* 11, no. 1 (2012): 12–13.

Deck, Andrew, and Vittoria Elliott. 'How Line Is Fighting Disinformation Without Sacrificing Privacy.' *Rest of World*, 7 March 2021. https://restofworld.org/2021/how-line-is-fighting-disinformation-without-sacrificing-privacy/.

Deck, Andrew, and William Yang. 'To Survive, China's Biggest Dating App Became a Pharmacy.' *Rest of World*, 27 August 2021. https://restofworld.org/2021/china-bluecity-healthcare-dating-apps/.

'Defining Emerging Markets.' *The Economist*, 15 October 2017. https://www.economist.com/special-report/2017/10/05/defining-emerging-markets.

'Defying Ban, Chinese Apps Quietly Grow in India, Again.' *Times of India*, 30 August 2021. https://timesofindia.indiatimes.com/business/india-business/defying-ban-chinese-apps-find-a-way-to-grow-in-indiaagain/articleshow/85745057.cms.

DeNardis, Laura. *The Internet in Everything: Freedom and Security in a World with No Off Switch*. New Haven, CT: Yale University Press, 2020.

Dencik, Lina, Arne Hintz, and Jonathan Cable. 'Towards Data Justice? The Ambiguity of Anti-Surveillance Resistance in Political Activism.' *Big Data & Society* 3, no. 2 (2016).

Denyer, Simon, and Annie Gowen. 'Too Many Men.' *Washington Post*, April 18, 2018. https://www.washingtonpost.com/graphics/2018/world/too-many-men/.

Devlin, Kate. 'Sex Robots: How Do Sex Robots Work and Can You Buy a Sex Robot?' *Cosmopolitan*, 12 July 2021. https://www.cosmopolitan.com/uk/love-sex/sex/a36480612/sex-robots/.

'The Digital Leisure Divide and the Forcibly Displaced: Part 1: Desk Research.' UNHCR: The UN Refugee Agency, April 2022. https://www.unhcr.org/innovation/wp-content/uploads/2022/05/The-Digital-Leisure-Divide-and-the-Forcibly-Displaced.pdf.

'The Digital Leisure Divide and the Forcibly Displaced, Part 2: Field Research.' UNHCR: The UN Refugee Agency, April 2023. https://www.unhcr.org/innovation/wp-content/uploads/2023/04/The-Digital-Leisure-Divide-Field-Research.pdf.

Döring, Nicola, M. Rohangis Mohseni, and Roberto Walter. 'Design, Use, and Effects of Sex Dolls and Sex Robots: Scoping Review.' *Journal of Medical Internet Research* 22, no. 7 (2020): e18551.

Drabinski, Emily. 'Ideologies of Boring Things: The Internet and Infrastructures of Race.' *Los Angeles Review of Books*, 13 February 2018. https://lareviewofbooks.org/article/ideologies-of-boring-things-the-internet-and-infrastructures-of-race/#.

Durkheim, Émile. 'Elementary Forms of Religious Life.' In *Social Theory Re-Wired: New Connections to Classical and Contemporary Perspectives*, edited by Wesley Longhofer and Daniel Winchester, 52–67. New York: Routledge, 2016.

Dutta, Sudeshna. 'National/Official Languages in India: Check All Official Languages.' *Edudwar*, January 10, 2023. https://www.edudwar.com/national-official-languages-in-india/.

Ehrlich, Paul R. 'The Population Bomb.' *New York Times*, 4 November 1970. https://www.nytimes.com/1970/11/04/archives/the-population-bomb.html.

El Habti, Hicham. 'Why Africa's Youth Hold the Key to Its Development Potential.' World Economic Forum, 19 September 2022. https://www.weforum.org/agenda/2022/09/why-africa-youth-key-development-potential/.

'Envious Suburban Lifestyle of Urban Angst.' *Global Times*, 4 June 2021. https://www.globaltimes.cn/page/202106/1225406.shtml.

Erdmann, Nicola, and Sophia Ankel. 'How the Design of iPhones, Toilets, and Crash Test Dummies Has Ignored Women—and Sometimes Put Their Lives at Risk.' *Business Insider*, 12 October 2019. https://www.businessinsider.com/design-details-show-the-world-been-built-for-men-2019-10?international=true&r=US&IR=T.

Ewe, Koh. 'We Asked the Math Tutor Who Posts His Lessons on Pornhub: Why?' *Vice*, 11 November 2021. https://www.vice.com/en/article/dyp8wa/math-tutor-video-lessons pornhub-changhsu.

'Facebook Is Nearing a Reputational Point of No Return.' *The Economist*, 9 October 2021. https://www.economist.com/leaders/2021/10/09/facebook-is-nearing-a-reputational-point-of-no-return.

Field, Hayden. 'Remote Employee Monitoring Tech Is Surging.' *Tech Brew*, November 23, 2022. https://www.emergingtechbrew.

com/stories/2022/11/23/remote-employee-monitoring-tech-is-surging.

Fight for the Future. '25+ Human Rights Organizations Demand Zoom Abandon Emotion Tracking and Analysis.' Fight for the Future, May 11, 2022. https://www.fightforthefuture.org/news/2022-05-10-25-human-rights-organizations-demand-zoom-abandon-emotion-tracking-and-analysis-ai.

Financial Times Editorial Board. 'America Is Struggling to Counter China's Intellectual Property Theft.' *Financial Times*, April 18, 2022. https://www.ft.com/content/1d13ab71-bffd-4d63-a0bf-9e9bdfc33c39.

Florida, Richard. *The Rise of the Creative Class*. New York: Basic Books, 2019.

Fuchs, Christian. *Digital Humanism: A Philosophy for 21st Century Digital Society*. Bingley, UK: Emerald Group Publishing, 2022.

'Gallup Global Emotions 2022.' Gallup, 2022. https://img.lalr.co/cms/2022/06/29185719/2022-Gallup-Global-Emotions-Report-2022_compressed.pdf.

Gengler, Amanda M. 'Emotions and Medical Decision-Making.' *Social Psychology Quarterly* 83, no. 2 (2020): 174–194.

George. 'Brazil Has More Influencers than Dentists.' *News Bulletin*, May 28, 2022. https://newsbulletin247.com/economy/104592.html.

Gergan, Mabel, Sara Smith, and Pavithra Vasudevan. 'Earth beyond Repair: Race and Apocalypse in Collective Imagination.' *Environment and Planning D: Society and Space* 38, no. 1 (2020): 91–110.

Gladwell, Malcolm. 'I Was a Stranger and You Welcomed Me.' In *Revisionist History*, produced by Pushkin Industries, podcast, MP3 audio, July 28, 2022. https://www.pushkin.fm/podcasts/revisionist-history/i-was-a-stranger-and-you-welcomed-me.

Glutting, Jana-Chin Rué, and Apoorva Shankar. 'Sustainability in the Global South Isn't One-Size-Fits-All—Here's Why.' *Circle*

Economy, November 22, 2022. https://www.circle-economy.com/blogs/sustainability-in-the-global-south-isnt-one-size-fits-all-heres-why.

González-González, Carina S., Rosa Gil-Iranzo, and Patricia Paderewski-Rodríguez. 'Sex with Robots: Analyzing the Gender and Ethics Approaches in Design.' In *Interacción '19: Proceedings of the XX International Conference on Human Computer Interaction* (2019): 1–8. https://doi.org/10.1145/3335595.3335609.

Green, Graeme. 'Five Ways AI Is Saving Wildlife—from Counting Chimps to Locating Whales.' *The Guardian*, February 21, 2022. https://www.theguardian.com/environment/2022/feb/21/five-ways-ai-is-saving-wildlife-from-counting-chimps-to-locating-whales-aoe.

Grimley, Naomi. 'Identity 2016: 'Global Citizenship' Rising, Poll Suggests.' *BBC News*, April 28, 2016. https://www.bbc.com/news/world-36139904.

Groenewald, Arline. 'Indonesia Blocks Popular Websites under New Censorship Law.' *Privacy Hub*, August 4, 2022. https://www.cyberghostvpn.com/en_US/privacyhub/indonesia-blocks-websites/.

Gul, Mahwish. 'Biological Facts Every Girl Must Know.' *D+C*, December 30, 2021. https://www.dandc.eu/en/article/new-digital-app-helps-break-taboo-surrounding-girls-menstruation-pakistan.

Guo, Eileen, and Adi Renaldi. 'Deception, Exploited Workers, and Cash Handouts: How Worldcoin Recruited Its First Half a Million Test Users.' *MIT Technology Review*, April 6, 2022. https://www.technologyreview.com/2022/04/06/1048981/worldcoin-cryptocurrency-biometrics-web3/.

Gurmat, Sabah. 'India Is Creating a 'Surveillance Segregation' State by Tracking Lower-Caste Workers.' *Daily Dot*, November 6, 2022. https://www.dailydot.com/debug/india-sanitation-workers-gps-watches-surveillance-segregation/.

Haberly, Daniel, and Dariusz Wójcik. *Sticky Power: Global Financial Networks in the World Economy*. Oxford: Oxford University Press, 2022.

Hagendorff, Thilo, Leonie N. Bossert, Yip Fai Tse, and Peter Singer. 'Speciesist Bias in AI: How AI Applications Perpetuate Discrimination and Unfair Outcomes against Animals.' *AI and Ethics* 3, no. 3 (August 2023): 717–734.

Hallanan, Lauren. 'Amazon Live Is Alibaba's Live-Streaming without the Good Bits.' *Forbes*, March 15, 2019. https://www.forbes.com/sites/laurenhallanan/2019/03/15/amazon-live-is-alibabas-live-streaming-without-the-good-bits/?sh=3761fb8d94ab.

Han, Xiao, and Giselinde Kuipers. 'Humour and TikTok Memes during the 2020 Pandemic Lockdown: Tensions of Gender and Care Faced by Chinese Mothers Working from Home.' *China Information* 35, no. 3 (2021): 393–419.

Hanleigh. 'The Future of Apps Is Lite: Lightening the Data Burden on Africa's Cost-Conscious Mobile Community.' *SME Tech Guru*, October 13, 2020. https://www.smetechguru.co.za/the-future-of-apps-is-lite-lightening-the-data-burden-on-africas-cost-conscious-mobile-community/.

Harper, Jennifer. 'Rhino Horn Now More Valuable than Gold, Diamonds and Cocaine, at $60K a Pound.' *Washington Times*, May 17, 2015. https://www.washingtontimes.com/news/2015/may/17/rhino-horn-considered-cure-all-and-aphrodesiac-now/.

Henrich, Joseph, Steven J. Heine, and Ara Norenzayan. 'Most People Are Not WEIRD.' *Nature* 466 (2010): 29.

Highmore, Ben. 'Bitter after Taste: Affect, Food and Social Aesthetics.' In *The Affect Theory Reader*, edited by Melissa Gregg and Gregory J. Seigworth, 118–200. Durham, NC: Duke University Press, 2010.

Hine, Emma, and Luciano Floridi. 'New Deepfake Regulations in China Are a Tool for Social Stability, but at What Cost?' *Nature Machine Intelligence* 4 (2022): 608–610. https://doi.org/10.1038/s42256-022-00513-4.

Hodgett, Susan. 'Sen, Culture and Expanding Participatory Capabilities in Northern Ireland.' *Journal of Human Development* 9, no. 2 (2008): 165–183.

Holler, Jacqueline. 'The History of Emotions in Colonial Latin America.' *Oxford Research Encyclopedia of Latin American History*, May 23, 2019. https://doi.org/10.1093/acrefore/9780199366439.013.533.

'How Asia Is Crucial in the Battle against Climate Change.' *The Economist*, October 27, 2021. https://www.economist.com/special-report/2021/10/27/how-asia-is-crucial-in-the-battle-against-climate-change.

'How the West Fell Out of Love with Economic Growth.' *The Economist*, December 11, 2022. https://www.economist.com/finance-and-economics/2022/12/11/how-the-west-fell-out-of-love-with-economic-growth.

Huddleston, Tom. 'Carly Fiorina Says the Chinese 'Don't Innovate.'' *Time*, May 26, 2015. https://time.com/3897081/carly-fiorina-china-innovation/.

'Insights into Latin America's $3.5 Billion Mobile Games Market: Players, Payers, Revenues, Esports & Market Dynamics.' Newzoo, July 8, 2021. https://newzoo.com/insights/articles/insights-into-latin-americas-3-5-billion-mobile-games-market-players-payers-revenues-esports-market-dynamics.

'Intel AI Protects Animals with National Geographic Society, Leonardo DiCaprio Foundation.' Intel Newsroom, January 13, 2019. https://intercontactnews.com/2019/01/13/intel-ai-protects-animals-with-national-geographic-society-leonardo-dicaprio-foundation/.

International Finance Corporation. *Climate Investment Opportunities in Emerging Markets: An IFC Analysis*. Washington, DC: International Finance Corporation, 2016. https://www.ifc.org/content/dam/ifc/doc/mgrt/3503-ifc-climate-investment-opportunity-report-dec-final.pdf.

Jacobs, Jane. *The Death and Life of Great American Cities*. New York: Vintage Books, 1993.

James, George Alfred. *Ecology Is Permanent Economy: The Activism and Environmental Philosophy of Sunderlal Bahuguna*. Albany: SUNY Press, 2013.

Jenkins, Beverly L. 'Monica Lewinsky Creates 'Goodness Bot' to Combat Mean Tweets and Cyberbullying.' *InspireMore*, December 4, 2019. https://www.inspiremore.com/monica-lewinsky-goodness-bot/.

Johar, Indy, and Kate Raworth. 'Design for Planet: The Big Picture.' Presented at the Design for Planet Festival, 2022. Vimeo video, 52:54. https://vimeo.com/657489496.

Kant, Immanuel. 'On the Use of Teleological Principles in Philosophy.' In *Anthropology, History, and Education*, edited by Günter Zöller, 192–218. Cambridge: Cambridge University Press, 1788.

Karger, Howard Jacob. 'The 'Poverty Tax' and America's Low-Income Households.' *Families in Society* 88, no. 3 (2007): 413–417.

Katwala, Amit. 'The Spiralling Environmental Cost of Our Lithium Battery Addiction.' *Wired*, August 5, 2018. https://www.wired.co.uk/article/lithium-batteries-environment-impact.

Kemp, Simon. 'Digital 2021: Global Overview Report.' *DataReportal*, January 27, 2021. https://datareportal.com/reports/digital-2021-global-overview-report.

Kerkelä, Julia. "Land Is Life': Climate Change Is Destroying the Arctic and the Sámi Way of Life.' *The Greens*, February 6, 2022.

https://www.greens-efa.eu/opinions/sami-climate-change-in-the-arctic/.

Khair, Tabish. 'Kipling on the Phantom Rickshaw: Between Words and a Hard Place.' *PN Review* 27, no. 1 (2009): 9–10.

Khalil, Abubakar Nur. 'The Nigerian Election and Naira Crisis Are Fueling Bitcoin Adoption.' *Forbes*, March 15, 2023. https://www.forbes.com/sites/abubakarnurkhalil/2023/03/01/nigerian-election-and-naira-crisis-is-fueling-bitcoin-adoption/?sh=296bb3cb35d4.

Khan, Saad. 'The Livestreaming App Connecting the Working Class.' *Rest of World*, December 21, 2020. https://restofworld.org/2020/app-connecting-south-asian-working-class/.

Khan, Zia. 'A Bretton Woods for AI: Ensuring Benefits for Everyone.' Rockefeller Foundation (blog), August 7, 2020. https://www.rockefellerfoundation.org/blog/a-bretton-woods-for-ai-ensuring-benefits-for-everyone/.

Kiaga, Annamarie, and Vicky Leung. 'The Transition from the Informal to the Formal Economy in Africa.' *Global Employment Policy Review* background paper no. 4, International Labour Organization, December 2020. https://www.ilo.org/wcmsp5/groups/public/---ed_emp/documents/publication/wcms_792078.pdf.

Kilgore, Georgette. 'Carbon Footprint of Data Centers & Data Storage per Country (Calculator).' 8 Billion Trees, March 28, 2023. https://8billiontrees.com/carbon-offsets-credits/carbon-ecological-footprint-calculators/carbon-footprint-of-data-centers/.

Klare, Michael T. 'The Future of AI Is War.' *The Nation*, July 17, 2023. https://www.thenation.com/article/world/artificial-intelligence-us-military/.

Klein, Matt. '50 Million Join the 'Creator Economy' Thanks to Platforms Like OnlyFans, YouTube, Etsy and Twitch.' *Forbes*,

September 23, 2020. https://www.forbes.com/sites/mattklein/2020/09/23/50m-join-the-creator-economy-as-new-platforms-emerge-to-help-anyone-produce-content--money/?sh=1d14aa33165f.

Klein, Naomi. *The Shock Doctrine: The Rise of Disaster Capitalism*. New York: Macmillan, 2007.

Klimek, Axel, and Alan AtKisson. *Parachuting Cats into Borneo: And Other Lessons from the Change Café*. White River Junction, VT: Chelsea Green Publishing, 2016.

Kohn, Eduardo. *How Forests Think: Toward an Anthropology beyond the Human*. Oakland: University of California Press, 2013.

Komarraju, Sai Amulya, Payal Arora, and Usha Raman. 'Agency and Servitude in Platform Labour: A Feminist Analysis of Blended Cultures.' *Media, Culture & Society* 44, no. 4 (2022): 672–689.

Kramer, Adam D. I., Jamie E. Guillory, and Jeffrey T. Hancock. 'Experimental Evidence of Massive-Scale Emotional Contagion through Social Networks.' *Proceedings of the National Academy of Sciences* 111, no. 24 (2014): 8788–8790.

Lake, Osprey Orielle. 'Women Rising for the Earth.' *Earth Island Journal* (Autumn 2018). https://www.earthisland.org/journal/index.php/magazine/entry/women-rising-for-the-earth/.

Ledgard, Jimena. 'Getting Political Is Becoming Profitable for Influencers in Peru.' *Rest of World*, March 12, 2021. https://restofworld.org/2021/getting-political-is-becoming-profitable-for-influencers-in-peru/.

Li, Fei-Fei. 'Opening the Google AI China Center.' *The Keyword* (blog), Google, December 13, 2017. https://blog.google/around-the-globe/google-asia/google-ai-china-center/.

'Life on the Frontline 2018: A Global Survey of the Working Conditions of Rangers.' World Wildlife Fund, October 9, 2018. https://www.worldwildlife.org/publications/life-on-the-frontline-2018.

Lin, Jean, and Jeroen de Kloet. 'Platformization of the Unlikely Creative Class: Kuaishou and Chinese Digital Cultural Production.' *Social Media & Society* 5, no. 4 (2019): 1–12. https://doi.org/10.1177/2056305119883430.

Lohr, Steve. 'Facial Recognition Is Accurate If You're a White Guy.' *New York Times*, February 9, 2018. https://www.nytimes.com/2018/02/09/technology/facial-recognition-race-artificial-intelligence.html.

Lovink, Geert. *Sad by Design: On Platform Nihilism*. London: Pluto Press, 2019.

Lovink, Geert. *Stuck on the Platform: Reclaiming the Internet*. Amsterdam: Valiz, 2022.

Lupton, Deborah. *The Internet of Animals: Human-Animal Relationships in the Digital Age*. New York: John Wiley & Sons, 2023.

Lynch, Leah, and Chung Tsz Chung. *E-commerce and Mobile Money for Poverty Reduction in China: Lessons for African Countries*. Washington, DC: Center for Global Development, 2020. https://www.cgdev.org/sites/default/files/e-commerce-and-mobile-money-poverty-reduction-china-lessons-african-countries.pdf.

Macaulay, Thomas Babington. 'Minute by the Hon'ble T. B. Macaulay, Dated the 2nd February 1835.' In *Selections from Educational Records, Part I (1781–1839)*, edited by H. Sharp. Calcutta: Superintendent, Government Printing, 1920. Reprint: Delhi: National Archives of India, 1965. http://www.columbia.edu/itc/mealac/pritchett/00generallinks/macaulay/txt_minute_education_1835.html.

Mangan, Lucy. 'Unknown: Killer Robots Review—the Future of AI Will Fill You with Unholy Terror.' *The Guardian*, July 10, 2023. https://www.theguardian.com/tv-and-radio/2023/jul/10/unknown-killer-robots-review-the-future-of-ai-will-fill-you-with-unholy-terror.

Marantz, Andrew. 'How 'Silicon Valley' Nails Silicon Valley.' *New Yorker*, June 9, 2016. https://www.newyorker.com/culture/culture-desk/how-silicon-valley-nails-silicon-valley.

Martinez-Alier, J. 'The Environment as a Luxury Good or 'Too Poor to Be Green'?' *Ecological Economics* 13, no. 1 (1995): 1–10.

Mashelkar, R. A. 'Gandhian Engineering: More from Less for More.' Lecture, Lal Bahadur Shastri Institute of Management. Accessed May 14, 2023. https://mashelkar.com/speeches/gandhian-engineering-more-from-less-for-more/.

Mattos, Agatha, Michela Bertolotto, and Gavin McArdle. 'Mapping Slums with Deep Learning Feature Extraction.' Paper presented at CDCEO 2022: 2nd Workshop on Complex Data Challenges in Earth Observation, Vienna, Austria, July 25, 2022. https://ceur-ws.org/Vol-3207/paper5.pdf.

Mazumdaru, Srinivas. 'Bretton Woods at 75: Has the System Reached Its Limits?' *Deutsche Welle*, July 22, 2019. https://www.dw.com/en/bretton-woods-at-75-has-the-system-reached-its-limits/a-49687599.

McTague, Tom. 'Yes, Make Coronavirus Jokes.' *The Atlantic*, April 3, 2020. https://www.theatlantic.com/international/archive/2020/04/humor-laughter-coronavirus-covid19/609184/.

'Meditation Apps—Worldwide.' Statista. Accessed May 14, 2023. https://www.statista.com/outlook/dmo/digital-health/digital-fitness-well-being/digital-fitness-well-being-apps/meditation-apps/worldwide#global-comparison.

Melnichuk, Tatiana. 'Why Does Emotional Intelligence Matter in Recruiting?' *Forbes*, October 18, 2021. https://www.forbes.com/sites/forbeshumanresourcescouncil/2021/10/18/why-does-emotional-intelligence-matter-in-recruiting/?sh=65e55bdb2f62.

Melton, Chad A., Brianna M. White, Robert L. Davis, Robert A. Bednarczyk, Arash Shaban-Nejad. 'Fine-Tuned Sentiment Analysis of COVID-19 Vaccine–Related Social Media Data: Comparative Study.' *Journal of Medical Internet Research* 24, no. 10 (2022): e40408.

Mendes, Lisa. "Africa Is the Future': All Eyes Turn to the Youngest Continent as the Next Frontier for Growth.' *Trade Finance Global*, August 24, 2022. https://www.tradefinanceglobal.com/posts/africa-is-the-future-all-eyes-turn-to-youngest-continent-as-next-frontier-for-growth/.

Mohandas, Shweta, and Deepika Nandagudi Srinivasa. 'The Boss Will See You Now: The Growth of Workplace Surveillance in India, Is Data Protection Legislation the Answer?' Centre for Internet and Society, December 31, 2020. https://cis-india.org/internet-governance/blog/the-boss-will-see-you-now-the-growth-of-workplace-surveillance-in-india-is-data-protection-legislation-the-answer.

Mpela, Mlungisi. 'Top 10 Countries That Watch the Most Porn!' *Sun Daily*, July 17, 2022. https://www.snl24.com/dailysun/lifestyle/top-10-countries-that-watch-the-most-porn-20220715.

Mulakala, Anthea. 'China's Global Development Initiative: Soft Power Play or Serious Commitment?' *DevPolicy Blog*, Development Policy Centre, Australian National University, October 18, 2022. https://devpolicy.org/chinas-gdi-soft-power-play-or-serious-commitment-20221018/.

Muradian, Roldan. 'Frugality as a Choice vs. Frugality as a Social Condition: Is De-growth Doomed to Be a Eurocentric Project?' *Ecological Economics* 161 (2019): 257–260.

Nascimento, Elen C. Carvalho, Eugênio da Silva, and Rodrigo Siqueira-Batista. 'The 'se' of Sex Robots: A Bioethical Issue.' *Asian Bioethics Review* 10, no. 3 (2018): 231–240.

'New Report from Richard Florida Aims to Define Size and Propel the Creator Economy.' Globe Newswire, November 1, 2022. https://www.globenewswire.com/en/news-release/2022/11/01/2545774/0/en/New-Report-From-Richard-Florida-Aims-to-Define-Size-and-Propel-the-Creator-Economy.html.

Noble, Safiya Umoja. *Algorithms of Oppression: How Search Engines Reinforce Racism*. New York: New York University Press, 2018.

Nodder, Chris. *Evil by Design: Interaction Design to Lead Us into Temptation*. Indianapolis, IN: John Wiley & Sons, 2013.

Norman, Don. *Design for a Better World: Meaningful, Sustainable, Humanity Centered*. Cambridge, MA: MIT Press, 2023.

Omifolaji, James Kehinde, Alice C. Hughes, Abubakar Sadiq Ibrahim, Jinfeng Zhou, Siyuan Zhang, Emmanuel Tersea Ikyaagba, and Xiaofeng Luan. 'Dissecting the Illegal Pangolin Trade in China: An Insight from Seizures Data Reports.' *Nature Conservation* 46, no. 17 (2022). https://natureconservation.pensoft.net/article/57962/.

Ong, Angelica Gustilo, and Eduardo Gustale. 'Frugal Innovations to Accelerate Sustainable Development.' *Accelerator Labs* (blog), United Nations Development Program, December 30, 2020. https://www.undp.org/acceleratorlabs/blog/frugal-innovations-accelerate-sustainable-development.

Onuoha, Mimi. 'Side-by-Side Images Expose a Glitch in Google's Maps.' *Quartz*, June 6, 2017. https://qz.com/982709/google-maps-is-making-entire-communities-invisible-the-consequences-are-worrying.

Orlowski-Yang, Jeff, dir. *The Social Dilemma*. Boulder, CO: Exposure Labs and Argent Pictures, 2020, 1:34:00. https://www.thesocialdilemma.com/.

Osterath, Brigitte. 'Why Life Thrives in the Foggy Namib Desert.' *Deutsche Welle*, August 14, 2017. https://www.dw.com/en/why-life-thrives-in-the-foggy-namib-desert/a-39622196.

'Owner of China's Biggest Gay Dating App Enters Online Healthcare Business Focusing on Men's Health.' *Global Times*, April 20, 2021. https://www.globaltimes.cn/page/202104/1221615.shtml.

Parashar, Priyanka. 'Female Workforce in India Declined to 9% Since Covid Pandemic. Here's Why.' Mint, June 3, 2022. https://www.livemint.com/news/india/female-workforce-in-india-declined-to-9-since-covid-pandemic-here-s-why-11654254827989.html.

Parater, Lauren. 'Beyond Destruction: Innovation as an Offering for Repair, Renewal and Reparations.' *Project Unsung*, Medium, February 11, 2022. https://medium.com/project-unsung/beyond-destruction-innovation-as-an-offering-for-repair-renewal-and-reparations-4a3c89293350.

Perez, Caroline Criado. *Invisible Women: Data Bias in a World Designed for Men*. Chicago: Abrams, 2019.

Perez, Sarah. 'Meta's NPE Team Takes a Global Focus with Seed-Stage Investments, Offices in Emerging Markets.' *TechCrunch*, December 2, 2021. https://techcrunch.com/2021/12/02/metas-npe-team-takes-a-global-focus-with-seed-stage-investments-offices-in-emerging-markets/.

Perez, Sarah. 'TikTok Launches Its First Creator Crediting Tool to Help Video Creators Cite Their Inspiration.' *TechCrunch*, May 18, 2022. https://techcrunch.com/2022/05/18/tiktok-launches-its-first-creator-crediting-tool-to-help-video-creators-cite-their-inspiration/.

Pickren, Graham. 'Political Ecologies of Electronic Waste: Uncertainty and Legitimacy in the Governance of E-waste Geographies.' *Environment and Planning A* 46, no. 1 (2014): 26–45.

Poonam, Snigdha, and Samarth Bansal. 'Inside the Chinese Dating Apps Exploiting the Loneliness of India's Men.' *Quartz*, March 13, 2020. https://qz.com/india/1811151/chinese-dating-apps-are-exploiting-loneliness-of-indias-men.

'Pope Discusses Ethics of Artificial Intelligence with Microsoft Chief.' *Reuters*, February 13, 2019. https://www.reuters.com/article/us-technology-microsoft-pope-idUSKCN1Q22B0.

Poushter, Jacob, and Nicholas Kent. 'The Global Divide on Homosexuality Persists.' Pew Research Center, June 25, 2020. https://www.pewresearch.org/global/2020/06/25/global-divide-on-homosexuality-persists/.

Prahalad, C. K., and Stuart L. Hart. *The Fortune at the Bottom of the Pyramid*. Philadelphia: Wharton School Publishing, 2005.

Rahman, Md. Shahinoor. 'A Foreigner's Perspective on Chinese Creativity and Innovation.' *China Daily*, January 30, 2019. https://www.chinadaily.com.cn/a/201901/30/WS5c516097a3106c65c34e75c8.html.

Rainsford, Sarah. 'Bitcoin Babushkas: Cryptocurrency Mining in Siberia.' *BBC News*, December 29, 2018. https://www.bbc.com/news/av/world-europe-46634596.

Rakshit, Devrupa. 'Why We Find Pessimists to Be More Intelligent.' *The Swaddle*, July 25, 2022. https://theswaddle.com/why-we-find-pessimists-to-be-more-intelligent/.

Ramachandran, Sreelakshmi, and Snehil Singh. 'Women in the Driving Seat: Studying the Interplay between Gender and Asset Ownership in the Platform Economy.' Ola Mobility Institute, March 2021. https://olawebcdn.com/ola-institute/women_asset_ownership_platform_economy.pdf.

Rangaswamy, Nimmi, and Payal Arora. 'The Mobile Internet in the Wild and Every Day: Digital Leisure in the Slums of Urban India.' *International Journal of Cultural Studies* 19, no. 6 (2016): 611–626.

Rashid, Atikh. 'Migrant Workers Gone, Pune Grapples with Shortage of Labourers; Work Speed Down to 20%.' *Indian Express*, June 1, 2020. https://indianexpress.com/article/cities/pune/effects-of-migrant-exodus-metro-sites-wear-desolate-look-progress-of-work-down-to-20-per-cent-of-usual-capacity-in-pune-6436392/.

Raworth, Kate. *Doughnut Economics: Seven Ways to Think Like a 21st-Century Economist*. White River Junction, VT: Chelsea Green, 2017.

Recio, Eugenia, and Dina Hestad. 'Still Only One Earth: Lessons from 50 Years of UN Sustainable Development Policy.' International Institute for Sustainable Development, April 22, 2022. https://www.iisd.org/articles/deep-dive/indigenous-peoples-defending-environment-all.

Reddy, William. *The Navigation of Feeling: A Framework for the History of Emotions*. New York: Cambridge University Press, 2001.

'Reliance JioMart and Meta to Enable Grocery Shopping on WhatsApp.' Retail Insight Network, August 30, 2022. https://www.retail-insight-network.com/news/jiomart-whatsapp-grocery-retail/.

Ridley, Matt. *The Rational Optimist: How Prosperity Evolves*. New York: Harper Perennial, 2010.

Rittel, Horst W. J., and Melvin M. Webber. 'Dilemmas in a General Theory of Planning.' *Policy Sciences* 4 (1973): 155–169. https://www.sympoetic.net/Managing_Complexity/complexity_files/1973%20Rittel%20and%20Webber%20Wicked%20Problems.pdf.

Ruiz, Michelle. 'Safiya Noble Knew the Algorithm Was Oppressive.' *Vogue*, October 21, 2021. https://www.voguc.com/article/safiya-noble.

Rusch, Natasha. "Intelligence That Will Win This War': Investigating Artificial Intelligence for Conservation.' Master's thesis, Erasmus University Rotterdam, 2020. https://thesis.eur.nl/pub/55368/Rusch-Natasha.pdf.

Rusch, Natasha, and Payal Arora. 'AI-Based Strategies to Combat Wildlife Trafficking and Wet Markets in Asia: A Critical Review.' In *The Next Digital Decade: Case Studies from Asia*. Vol. 1, *Traces and*

Divides, edited by Dev Lewis, Katharina Naumann, and Ming Yin Ho. Singapore and Hong Kong: Konrad-Adenauer-Stiftung / Digital Asia Hub, 2021. https://www.digitalasiahub.org/publications/next-digital-decade-case-studies-from-asia-vol-1-traces-and-divides/.

Sampath, G. 'Is Kissing in Public an Obscene Act?' Mint, November 4, 2014. https://www.livemint.com/Opinion/5DKL9wuITBg8LosTQFFX8O/Is-kissing-in-public-an-obscene-act.html.

'Sanitation Workers Protest against GPS Enabled Watches, Chandigarh MC Refuses to Budge.' *Hindustan Times*, March 13, 2020. https://www.hindustantimes.com/chandigarh/sanitation-workers-protest-against-gps-enabled-watches-chandigarh-mc-refuses-to-budge/story-sYJLH96Yi6smfctGtz0BII.html.

Seaman, Olivia. 'How TikTok's Body Positivity Movement Overshadows Body Shaming.' *The Torch*, December 14, 2021. https://www.torchonline.com/opinion/2021/12/14/how-tiktoks-body-positivity-movement-overshadows-body-shaming/.

'Secret App Banned in Brazil, but Loopholes Exist.' International Association of Privacy Professionals (IAPP). Last modified August 26, 2014. https://iapp.org/news/a/secret-app-banned-in-brazil-but-loopholes-exist/.

Service Civil International. 'Picturing the Global South: The Power behind Good Intentions.' Version 3. Service Civil International Toolkit, 2018. http://www.sci.or.at/wp-content/uploads/2016/05/picturingtheglobalsouth_toolkit_v1.pdf.

'Shampoo Sachets Are Not Plastic as They Appear.' News, The Fletcher School, Tufts University, December 14, 2022. https://fletcher.tufts.edu/news-events/news/shampoo-sachets-are-not-plastic-they-appear.

Sharma, Achyutha. 'Women Resellers in India's Gig Economy: From Access to Confidence.' In *Feminist Futures of Work: Reimagining Labour in the Digital Economy*, edited by Payal Arora, Usha Raman, and René König, 77–86. Amsterdam: Amsterdam University Press, 2023. https://doi.org/10.2307/jj.2711713.10.

Sharma, Niharika. 'Reliance Jio's Cheap Data Turned India's Internet Dreams into Reality.' *Quartz*, September 7, 2021. https://qz.com/india/2055771/reliance-jios-cheap-data-turned-indias-internet-dreams-into-reality.

Shiva, Vandana. *Monocultures of the Mind: Perspectives on Biodiversity and Biotechnology*. London: Zed Books, 1993.

Shiva, Vandana. *The Violence of the Green Revolution: Third World Agriculture, Ecology and Politics*. London: Zed Books, 1991.

Sias, Leonardo. 'The Ideology of AI.' *Philosophy Today* 65, no. 3 (2012): 505–522.

Silberling, Amanda. 'Facebook Whistleblower Frances Haugen Testifies before the Senate.' *TechCrunch*, October 5, 2021. https://techcrunch.com/2021/10/05/facebook-whistleblower-frances-haugen-testifies-before-the-senate/.

Snow, Jackie. 'How Africa Is Seizing an AI Opportunity.' *Fast Company*, October 3, 2019. https://www.fastcompany.com/90308114/how-africa-is-seizing-an-ai-opportunity.

Srinivasa, Bala. 'CavinKare—Shampoo Sachet That Was the True Pioneer of Middle India Market.' *The Guardian*, November 12, 2022. https://www.theguardian.com/sustainable-business/sachet-packaging-low-income-communities-waste-nightmare.

Srinivasan, Ramya, and Beatriz San Miguel González. 'The Role of Empathy for Artificial Intelligence Accountability.' *Journal of Responsible Technology* 9 (2022): 100021. https://www.sciencedirect.com/science/article/pii/S2666659621000147?via%3Dihub.

Stafford, Tom. 'Psychology: Why Bad News Dominates the Headlines.' *BBC News*, July 28, 2014. https://www.bbc.com/future/article/20140728-why-is-all-the-news-bad.

Stassen, Murray. 'Why Anghami, with over 1M Paying Users, Doesn't Feel Threatened by Spotify in the Middle East.' *Music Business Worldwide*, March 5, 2019. https://www.musicbusinessworldwide.com/anghami-with-over-1m-paying-users-doesnt-feel-threatened-by-spotify-in-the-middle-east/.

'The State of Conservation Technology.' WildLabs, December 2021. https://wildlabs.net/state-of-conservation-technology.

Stokel-Walker, Chris. *YouTubers: How YouTube Shook Up TV and Created a New Generation of Stars*. Surrey: Canbury Press, 2019.

Stoler, Ann Laura. 'Affective States.' In *A Companion to the Anthropology of Politics*, edited by David Nugent and Joan Vincent, 4–20. Malden, MA: Blackwell, 2004.

Swinnen, Lucy, Jack Goodman, Hannah Gelbart, Maria Korenyuk, and Juliana Gragnani. 'Telegram: Where Women's Nudes Are Shared without Consent.' *BBC News*, February 16, 2022. https://www.bbc.com/news/world-60303769.

TeamAG. 'POACHERS—the People behind the Statistics—a Reality Check.' *Africa Geographic*, October 21, 2020. https://africageographic.com/stories/poachers-the-people-behind-the-statistics-a-reality-check/.

'Tencent-Backed Kwai App Ranked Most Popular Social Short Video App.' *Business Insider*, May 28, 2018. https://markets.businessinsider.com/news/stocks/tencent-backed-kwai-app-ranked-most-popular-social-short-video-app-1025560745.

Thompson, Mark. 'Global Banking Crisis: What Just Happened?' CNN, March 20, 2023. https://edition.cnn.com/2023/03/17/business/global-banking-crisis-explained/index.html.

Tobin, Meaghan. 'Why China's Crypto Cowboys Are Fleeing to Texas.' *Rest of World*, August 11, 2021. https://restofworld.org/2021/bitcoin-miners-china-texas/.

Turner, Patricia. 'Why African-Americans Loathe 'Uncle Tom.'' *In Character*, NPR, July 30, 2008. https://www.npr.org/templates/story/story.php?storyId=93059468.

United Nations. 'SDG Indicators.' UN Statistics Division. Accessed November 21, 2023. https://unstats.un.org/sdgs/indicators/indicators-list/.

Vallance, Chris. 'Artificial Intelligence Could Lead to Extinction, Experts Warn.' *BBC News*, May 30, 2023. https://www.bbc.com/news/uk-65746524.

Vincent, James. 'AI Is Worse at Identifying Household Items from Lower-Income Countries.' *The Verge*, June 11, 2019. https://www.theverge.com/2019/6/11/18661128/ai-object-recognition-algorithms-bias-worse-household-items-lower-income-countries.

Vincent, James. 'Microsoft to Retire Controversial Facial Recognition Tool That Claims to Identify Emotion.' *The Verge*, June 21, 2022. https://www.theverge.com/2022/6/21/23177016/microsoft-retires-emotion-recognition-azure-ai-tool-api.

Wadhwa, Vivek, Ismail Amla, and Alex Sakever. 'As Silicon Valley Fantasizes about Web3, India Leaps Ahead on Payments.' *Fortune*, June 30, 2022. https://fortune.com/2022/06/30/blockchain-india-upi-payments-global-reach/.

Wadhwa, Vivek, Alex Salkever, and Ismail Amla. *From Incremental to Exponential: How Large Companies Can See the Future and Rethink Innovation*. Oakland, CA: Berrett-Koehler Publishers, 2020.

Wang, Xiaowei. *Blockchain Chicken Farm: And Other Stories of Tech in China's Countryside*. New York: FSG Originals x Logic, 2020.

Weitzberg, Kehren, Margie Chessman, Aaron Martin, and Emrys Schoemaker. 'Between Surveillance and Recognition:

Rethinking Digital Identity in Aid.' *Big Data & Society* 8, no. 1 (2021): 1–7.

'What if AI Were Developed to Serve Humanity rather than Commerce.' About Us, AI for Good. Accessed May 14, 2023. https://ai4good.org/about-us/.

Whitford, Emma. 'While #LearnOnTikTok Goes Viral, So Too Does Misinformation. McGraw Hill Is Fighting Back with Its Own App That Looks Like a 'Textbook and TikTok Had a Baby."' *Forbes*, October 18, 2022. https://www.forbes.com/sites/emmawhitford/2022/10/18/students-viewed-this-type-of-tiktok-412-billion-times-and-its-not-porn/?sh=663349407e9c.

'Why Is the EU's Right to Repair Legislation Important?' News, European Parliament, April 4, 2022, updated November 23, 2023. https://www.europarl.europa.eu/news/en/headlines/society/20220331STO26410/why-is-the-eu-s-right-to-repair-legislation-important.

'Why Protecting Indigenous Communities Can Also Help Save the Earth.' *The Guardian*, October 12, 2020. https://www.theguardian.com/climate-academy/2020/oct/12/indigenous-communities-protect-biodiversity-curb-climate-crisis.

Wolters, Claire. 'Asia's Live-Streaming Industry Promises Intimacy. So Why Are Users So Lonely?' *National Geographic*, July 31, 2019. https://www.nationalgeographic.com/culture/article/asia-live-streaming-industry-creates-loneliness.

Wood, David Murakami, and Rodrigo José Firmino. 'Empowerment or Repression? Opening Up Questions of Identification and Surveillance in Brazil through a Case of 'Identity Fraud."' *Identity in the Information Society* 2, no. 3 (2009): 299.

'The World's Most Valuable Resource Is No Longer Oil, but Data.' *The Economist*, May 6, 2017. https://www.economist.com/

leaders/2017/05/06/the-worlds-most-valuable-resource-is-no-longer-oil-but-data.

Wrigley, Sarah, and Toby Duthie. 'The Illegal Wildlife Trade Is Still One of the Most Profitable Criminal Enterprises.' *FCPA Blog*, April 27, 2020. https://fcpablog.com/2020/04/27/the-illegal-wildlife-trade-is-still-one-of-the-most-profitable-criminal-enterprises/.

Wu, Jun. 'Empathy in Artificial Intelligence.' *Forbes*, December 17, 2019. https://www.forbes.com/sites/cognitiveworld/2019/12/17/empathy-in-artificial-intelligence/?sh=3b8bfa363270.

Xu, Tony. 'The Stars Who Vanished from China's Live-Streaming Galaxy.' *TechNode*, April 15, 2019. https://technode.com/2019/04/15/the-stars-who-vanished-from-chinas-live-streaming-galaxy/.

y Tao, Andres Manner. 'What Do Youth Say about Higher Education in the Future?' UNESCO International Institute for Higher Education in Latin America and the Caribbean, September 26, 2022. https://www.iesalc.unesco.org/en/2022/09/26/what-do-youth-say-about-higher-education-in-the-future/.

Zakaria, Mohammed. 'Cops Tell Titillating Thai Street Vendor to Cover Up, Internet Creeps over Her Pics.' *Asia One*, November 26, 2021. https://www.asiaone.com/asia/cops-tell-titillating-thai-street-vendor-cover-internet-creeps-over-her pics.

Zhan, Echo. 'Li Ziqi Breaks YouTube Subscribers Record for Chinese Language Channel.' *Guinness World Records*, February 3, 2021. https://www.guinnessworldrecords.com/news/2021/2/li-ziqi-breaks-youtube-subscribers-record-for-chinese-language-channel-647202.

Zhang, Charlie Yi. *Dreadful Desires: The Uses of Love in Neoliberal China*. Durham, NC: Duke University Press, 2022.

Zhang Xiaoyu. 'China-India Cooperation in Digital Economy Has Great Potential but with Conditions.' *Global Times*, January 31, 2022. https://www.globaltimes.cn/page/202201/1250262.shtml.

Zhenhuan, Ma. 'Dance with Zhejiang Couple Lifts Spirits of Millions.' *China Daily*, June 8, 2020. https://global.chinadaily.com.cn/a/202006/08/WS5ede16e7a3108348172519a3.html.

Zhou, Viola. 'This Chinese Social Network Thinks the Key to Quality Conversation Is Anonymity.' *Rest of World*, August 30, 2022. https://restofworld.org/2022/3-minutes-with-zhang-lu-soul/.

Zorthian, Julia. 'OpenAI CEO Sam Altman Asks Congress to Regulate AI.' *Time*, May 16, 2023. https://time.com/6280372/sam-altman-chatgpt-regulate-ai/.

Zuboff, Shoshana. *The Age of Surveillance Capitalism: The Fight for a Human Future at the New Frontier of Power*. London: Profile Books, 2019.

Zygy, Sy. 'How a Carbon-Backed Cryptocurrency Is Tackling Climate Change.' World Economic Forum, July 12, 2022. https://www.weforum.org/agenda/2022/07/carbon-backed-cryptocurrency-tackling-climate-change/.

Notes

Scan this QR code to access the notes.

Index

About the Author

Payal Arora is a professor of Inclusive AI Cultures at Utrecht University. She is a digital anthropologist, a speaker and the author of award-winning books including *The Next Billion Users* (2019). Described as the 'next billion champion' by Forbes, Arora is sought after for her insights on inclusive design and policies based on her decades of user-experiences among marginalized communities worldwide. She is a Rockefeller Bellagio Resident Fellow Alumnus.

She has consulted for numerous public and private organizations, including UNHCR, UNESCO, Spotify, KPMG, Adobe and GE. She is Indian, American and Irish, and is currently based in Amsterdam.

HarperCollins *Publishers* India

At HarperCollins India, we believe in telling the best stories and finding the widest readership for our books in every format possible. We started publishing in 1992; a great deal has changed since then, but what has remained constant is the passion with which our authors write their books, the love with which readers receive them, and the sheer joy and excitement that we as publishers feel in being a part of the publishing process.

Over the years, we've had the pleasure of publishing some of the finest writing from the subcontinent and around the world, including several award-winning titles and some of the biggest bestsellers in India's publishing history. But nothing has meant more to us than the fact that millions of people have read the books we published, and that somewhere, a book of ours might have made a difference.

As we look to the future, we go back to that one word—a word which has been a driving force for us all these years.

Read.